Dirk Sanders

Ants and Spiders in Grassland Food Webs

Dirk Sanders

Ants and Spiders in Grassland Food Webs

Top-down control and intraguild interactions

Südwestdeutscher Verlag für Hochschulschriften

Impressum/Imprint (nur für Deutschland/ only for Germany)
Bibliografische Information der Deutschen Nationalbibliothek: Die Deutsche Nationalbibliothek verzeichnet diese Publikation in der Deutschen Nationalbibliografie; detaillierte bibliografische Daten sind im Internet über http://dnb.d-nb.de abrufbar.

Alle in diesem Buch genannten Marken und Produktnamen unterliegen warenzeichen-, marken- oder patentrechtlichem Schutz bzw. sind Warenzeichen oder eingetragene Warenzeichen der jeweiligen Inhaber. Die Wiedergabe von Marken, Produktnamen, Gebrauchsnamen, Handelsnamen, Warenbezeichnungen u.s.w. in diesem Werk berechtigt auch ohne besondere Kennzeichnung nicht zu der Annahme, dass solche Namen im Sinne der Warenzeichen- und Markenschutzgesetzgebung als frei zu betrachten wären und daher von jedermann benutzt werden dürften.

Verlag: Südwestdeutscher Verlag für Hochschulschriften Aktiengesellschaft & Co. KG
Dudweiler Landstr. 99, 66123 Saarbrücken, Deutschland
Telefon +49 681 37 20 271-1, Telefax +49 681 37 20 271-0, Email: info@svh-verlag.de
Zugl.: Göttingen, Universität Göttingen, Dissertation, 2008

Herstellung in Deutschland:
Schaltungsdienst Lange o.H.G., Berlin
Books on Demand GmbH, Norderstedt
Reha GmbH, Saarbrücken
Amazon Distribution GmbH, Leipzig
ISBN: 978-3-8381-0241-2

Imprint (only for USA, GB)
Bibliographic information published by the Deutsche Nationalbibliothek: The Deutsche Nationalbibliothek lists this publication in the Deutsche Nationalbibliografie; detailed bibliographic data are available in the Internet at http://dnb.d-nb.de.

Any brand names and product names mentioned in this book are subject to trademark, brand or patent protection and are trademarks or registered trademarks of their respective holders. The use of brand names, product names, common names, trade names, product descriptions etc. even without a particular marking in this works is in no way to be construed to mean that such names may be regarded as unrestricted in respect of trademark and brand protection legislation and could thus be used by anyone.

Publisher:
Südwestdeutscher Verlag für Hochschulschriften Aktiengesellschaft & Co. KG
Dudweiler Landstr. 99, 66123 Saarbrücken, Germany
Phone +49 681 37 20 271-1, Fax +49 681 37 20 271-0, Email: info@svh-verlag.de

Copyright © 2009 by the author and Südwestdeutscher Verlag für Hochschulschriften Aktiengesellschaft & Co. KG and licensors
All rights reserved. Saarbrücken 2009

Printed in the U.S.A.
Printed in the U.K. by (see last page)
ISBN: 978-3-8381-0241-2

Table of contents

General Introduction 5
   Generalist predators 6
   Top-down control and trophic cascades 6
   Intraguild predation 7
   Diversity of predators and ecosystem functioning 7
   Field experiments 8
   Stable isotopes 9
   References 10

Intraguild interactions between spiders and ants and top-down control in a grassland food web 15
   Abstract 16
   Introduction 17
   Materials and methods 17
   Results 20
   Discussion 26
   References 29

Predator diversity and top down effects: Intraguild interactions with hunting spiders reduce top-down control of web-builders 35
   Abstract 36
   Introduction 37
   Materials and methods 38
   Results 40
   Discussion 44
   References 46

Test for effects of functional diversity: ants, hunting spiders and web-builders in a wet grassland food web 51
   Abstract 52
   Materials and methods 54
   Results 55
   Discussion 60
   References 63

Habitat structure mediates top-down effects of spiders and ants on herbivores 67
   Abstract 68
   Introduction 69
   Material and methods 70
   Results 72
   Discussion 75
   References 77

Small scale habitat fragmentation affects generalist predator diversity: Implications for top-down control of spiders 81
   Abstract 82
   Introduction 83
   Material and methods 83
   Results 85
   Discussion 88
   References 90

Potential positive effect of the ant species *Lasius niger* on linyphiid spiders 95
   Abstract 96
   Introduction 97
   Material and methods 97
   Results 99
   Discussion 101

References 102
Conclusion 105
Food web interactions of generalist predators 106
Top-down control of generalist predators on herbivores and detritivores in natural grassland systems 108
Short answers to the questions I addressed in the introduction 109
References 109
Danksagung 111
Appendix 114

Chapter 1

4

Chapter 1

General Introduction

Generalist predators

Generalist predators have been thought to be poor biocontrol agents (Riechert and Lockeley 1984). This prediction was largely based on the lack of prey specificity and longer generation times than pests. On the other hand polyphagous predators, such as spiders, carabids, staphylinids and ants occur in large numbers in most terrestrial ecosystems, independently of specific prey populations (Ekschmitt *et al.* 1997). This pattern contrasts to that of specialized parasites and predators which colonize fields in response to a rising population density of their prey species. Because of their different life cycles generalist predators are present in different developmental stages throughout the year; some species are even active in winter. Thus, generalist predators seem to be ideal lying-in-wait predators because of their generalist feeding strategy. In this study I focused on spiders and ants as generalist predators because they are present in high densities in most terrestrial ecosystems (Wise 1993; Hölldobler and Wilson 1995), and indeed there were only very few samples from our studied grassland systems containing no spiders and ants. Spiders are known to be able to exert strong top-down control on herbivore populations as demonstrated in various experiments (Riechert & Bishop 1990; Riechert & Lawrence 1997; Schmitz 1998; Finke & Denno 2003; Cronin, Haynes, Dillemuth 2004) and contribute to the control of pest species in agricultural systems (Symondson *et al.* 2002; Lang 2003; Schmidt *et al.* 2003). Spiders can be classified into three major functional groups, according to their strategies for catching prey. Web-builders belonging to various families employ silk to assist in the capture of prey. The majority of web-building spiders at our study sites belong to the Linyphiidae, Araneidae and Theridiidae. Among the wandering spiders some lie motionless in ambush and are typical sit-and-wait predators (e.g. Thomisidae). Others are hunters, having in common that they actively go in search for their prey (e.g. Lycosidae, Pisaura, Salticidae). In contrast to spiders, most Central European ant species are omnivores, being able to prey on a wide range of other invertebrates, as well as to take up nutrients from plants indirectly by trophobiosis with phloem-feeding insects (Seifert 2007). Although they are omnivores, ants of the genus *Myrmica* and *Formica* can strongly affect the arthropod community. In their study of the role of *Myrmica* in a meadow ecosystem, Kajak *et al.* (1972) reported high predation rates of ants on juvenile arthropods, especially in the first half of the vegetation period. However, these effects were not demonstrated by field experiment but by observation and calculation. Another example for the top down control of ants is the reduction of foliage damage caused by moth during outbreaks by 34% in the presence of *Formica aquilonia* (Karhu 1998). Ants are also able to prey on large arthropods by recruiting nest mates, which largely extends the range of possible prey.

Top-down control and trophic cascades

Trophic cascades are indirect effects that are triggered by a direct effect of a predator (donor) on its prey (transmitter) and translated into changes in the prey's energy supply (receiver) in an interaction chain (Fig.1, Halaj & Wise 2001). Recent meta-analysis studies have shown that top-down forces by invertebrate predators on their herbivorous prey and cascade effects on plants play an important role in structuring communities in terrestrial ecosystems (Schmitz et al. 2000; Halaj & Wise 2001), although such effects are generally stronger in aquatic systems (Shurin et al. 2002). The actual role of trophic cascades in shaping the structure of terrestrial systems has been debated vigorously. Proponents of "the word is green hypothesis" (Hairston et al. 1960; Slobodkin et al. 1967) argue for the paramount role of predation. The hypothesis predicts that food-limited predators suppress herbivore population to an extent that herbivory is relatively unimportant. Strong top-down control in terrestrial ecosystems was mostly demonstrated for simply structured communities (e.g. Finke & Denno 2003; 2004; Schmidt et al. 2003). In food webs with a diverse species assemblage, top-down effects are thought to attenuate (Polis & Strong 1996; Polis 1999; Schmitz et al. 2000).

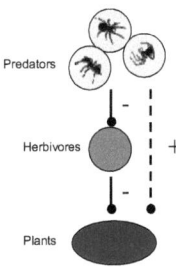

Fig 1. Trophic cascade generated by generalist predators

Intraguild predation

Intraguild predation, defined as feeding on species of the same guild, has gained relevance in ecological studies since Polis et al. (1989) pointed out its ecological and evolutionary implications. Organisms, that use the same resources in a similar way can be assigned to a certain guild. Hence, by preying on a member of the same guild the intraguild predator kills a potential competitor regarding their shared food resource. If the predators attack each other, the ultimate result is often relaxed predation pressure and diminished top-down control of shared prey (Fig. 2; Rosenheim et al. 1995, Snyder and Ives 2001, Finke and Denno 2003). Intraguild predation has been characterised as an important feature structuring arthropod communities (Wise 1993, Arim and Marquet 2004), especially if spiders are included. Spiders and ants are potential competitors and mutual predators. The ant species *Myrmica* seemed to exert a high predatory impact on spiders in a meadow (Petal and Breymeyer 1969; Kajak et al. 1972). Halaj et al. (1997) and Lenoir (2003) found negative effects of ants on

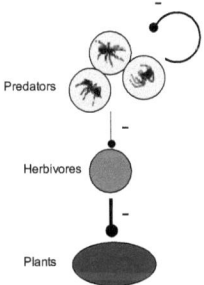

Fig 2. Intraguild predation reduces top down effects

the activity of spiders. However, in many studies such effects could not be demonstrated (Otto 1965; van der Aart and de Wit 1971; Lenoir et al. 2003; Gibb 2003). Similarly, Brüning (1991) tested the effects of *Formica polyctena* on spiders in a forest ecosystem without finding any evidence for intraguild predation.

Diversity of predators and ecosystem functioning

Declining biodiversity and its implications for continued provision of ecosystem services have led to an intense research effort to study the relationships between biodiversity and ecosystem functioning (Loreau et al. 2001, Wilby & Thomas 2002, Duffy 2003). While community ecology has historically focused on how ecological processes maintain species diversity, the central question of biodiversity and ecosystem functioning is how diversity affects, rather than responds to, ecological processes (Ives et al. 2005). Predators can strongly control herbivore populations, which can be an important ecosystem service regarding agricultural systems. Unfortunately, predators are more susceptible to extinctions than species at other trophic levels (Duffy 2002, 2003) and a change in the diversity of predators is known to affect the strength of trophic cascades (Finke & Denno 2004, Snyder et al. 2006). Increasing the species richness of plant, herbivore and filterfeeder communities has been shown to lead to a more efficient resource use at the community level (Tilman et al. 2001, Cardinale et al. 2002, Duffey 2003). There are two mechanisms thought to lead to this improvement in resource exploitation. First, communities including more species are more likely to include particularly efficient resource extractors by chance alone (sampling effect). The second mechanism is species complementarity, wherein species utilize resources in different ways such that total resource extraction is more complete within diverse communities (Snyder et al. 2006). Intraguild interactions can cause increasing predator diversity to disrupt the suppression of a pest species, thus leading to higher pest densities (Rosenheim et al. 1993, Finke and Denno 2004).

Field experiments

Experimental ecology lies within the spectrum of ecological methodology, a continuum that runs from passive observation to pure thought. Because patterns derived from passive observation cannot identify mechanisms whereas theories employing different mechanisms can describe the same pattern, an important objective in experimental ecology is to conduct a series of experiments that together provide sufficient information to link ecological theory with observation of nature (Fig. 3, Naem 2001). The ecological scale (spatial, temporal, and biotic scales) and the experimental validity are the two critical factors in designing experiments. Microcosm experiments under controlled conditions have a low external but a high internal validity, clearly demonstrating the effect of a factor, in contrast a field experiment under natural conditions has a high external but a low internal validity. External validity concerns the extent to which the results of an experiment can be generalized. Microcosm experiments are often used to study predation effects, however, they have serious limitations. The size and duration of microcosm experiments exclude or distort important features of natural communities and ecosystems and within the context of appropriately scaled field studies, microcosm experiments become irrelevant and diversionary (Carpenter 1996). Therefore, field experiments are an

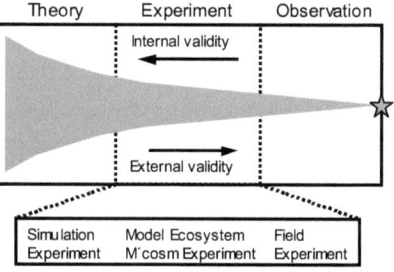

Fig 3. Experimental ecology within the theory-observation continuum. The star represents a precisely documented pattern in nature. The shaded area flaring out as one moves left represents the region in which ecologists work, the width of the band representing the range of possible studies. (Fig. from Naem 2001)

important method for studying trophic interactions and predatory effects under natural conditions (Wise 1993, Hodge 1999), providing a high external validity which is also based on the habitat chosen.

Stable isotopes

Additionally to the field experiments, I used the analysis of stable isotopes to gain a better understanding of trophic links in the particular food webs. This analysis provides important information for the discussion of predator effects observed in the field experiments. Stable isotope analysis of ratios of $^{15}N/^{14}N$ and $^{13}C/^{12}C$ is a promising tool for food web studies (De Niro and Epstein 1981; Wada et al. 1991; Kling et al 1992; Ponsard and Arditi 2000). Isotopic contents are expressed in δ units as the relative difference between sample and conventional standards with $\delta^{15}N$ or $\delta^{13}C$ [‰] = ($R_{Sample} - R_{Standard}$)/$R_{Standard}$ x 1000, where R is the ratio of $^{15}N/^{14}N$ or $^{13}C/^{12}C$ content, respectively. The conventional standard for ^{15}N is atmospheric nitrogen and for ^{13}C PD-belemnite (PDB) carbonate (Ponsard and Arditi 2000). Values of δ ^{13}C are largely conserved in food chains and provide information about the identity of the resource base (DeNiro and Epstein 1978; Petelle et al. 1979; Magnusson et al. 1999; Vander Zanden and Rasmussen 1999), whereas $\delta^{15}N$ values can be used as a trophic level indicator (Fig. 4; Ponsard and Arditi 2000; Post 2002; Vanderklift and Ponsard 2003). On average, the $^{15}N/^{14}N$ ratio of predators is increased by 3–4‰ compared with their prey (DeNiro and Epstein 1981; Minagawa and Wada 1984; Owens 1987; Peterson and Fry 1987; Cabana and Rasmussen 1994). However, within this general pattern variation in consumer diet $\delta^{15}N$ enrichment can be substantial (Vanderklift and Ponsard 2003).

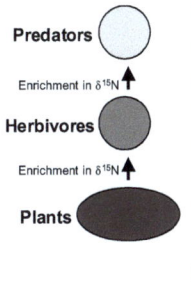

Fig 4 The $^{15}N/^{14}N$ ratio increases with the trophic level.

In this study I want to address the following questions, which were studied by using field experiments and stable isotopes.

(1) Can field experiments demonstrate the top-down control of generalist predators on herbivores in natural grassland systems?
(2) How important is intraguild predation regarding a predator guild that contain spiders and ants?
(3) Does a more diverse predator guild exert a stronger top-down control or do intraguild interactions reduce such effects?
(4) Do top-down effects attenuate with increasing structural complexity of habitat by providing refuges for herbivores?
(5) Habitat fragmentation is a common phenomenon in our landscape. Does small scale fragmentation affect the diversity and abundance of generalist predators?

References

Arim, M. and Marquet, P.A. (2004) Intraguild predation: a widespread interaction related to species biology. Ecology Letters, 7, 557–564.

Brüning A. (1991) The effect of a single colony of the red wood ant, *Formica polyctena*, on the spider fauna (Araneae) of a beech forest floor. Oecologia 86: 478-483.

Cabana G, Rasmussen JB (1994) Modelling food chain structure and contaminant bioaccumulation using stable nitrogen isotopes. Nature 372:255–257.

Cardinale, B.J., Palmer M.A., and Collins S.L. (2002) Species diversity enhances ecosystem functioning through interspecific facilitation. Nature 415:426-429.

Carpenter, S.R. (1996) Microcosm experiments have limited relevance for community ecosystem ecology. Ecology 77:677-680.

Cronin, J.T., Haynes, K.J., and Dillemuth, F. (2004) Spider effects on planthopper mortality, dispersal, and spatial popular dynamics. Ecology 85:2134-2143.

DeNiro, M.J., Epstein, S. (1978) Influence of diet on the distribution of carbon isotopes in animals. Geochim Cosmochim Acta 42:495–506.

DeNiro, M.J., Epstein, S. (1981) Influence of diet on the distribution of nitrogen isotopes in animals. Geochim Cosmochim Acta 45:341–351.

Duffy, J.E. (2002) Biodiversity and ecosystem function: the consumes connection. Oikos, 99, 201–219.

Duffy, J.E. (2003) Biodiversity loss, trophic skew and ecosystem functioning. Ecology Letters, 6, 680–687.

Ekschmitt, K., Weber, M., Wolters, V. (1997): Spiders, Carabids and Staphylinids - the ecological potential of predatory macroarthropods. In: Benckiser (Ed.): Fauna in Soil Ecosystems. Marcel Dekker, Inc. New York, pp. 307-362.

Finke, D.L., and Denno, R.F. (2003) Intra-guild predation relaxes natural enemy impacts on herbivore populations. Ecological Entomology 28: 67-73.

Finke, D.L., and Denno, R.F. (2004) Predator diversity dampens trophic cascades. Nature 429: 407-410.

Gibb, H. (2003) Dominant meat ants affect only their specialist predator in an epigaeic arthropod community. Oecologia 136:609-615.

Hairston, N.G., Smith F.E. and Slobodkin, L.B. (1960) Community structure, population control, and competition. American Naturalist 94:421-425.

Halaj, J., and Wise, D.H. (2001). Terrestrial trophic cascades: How much do they trickle? American Naturalist 157: 262-281.

Halaj, J., Ross, D.W. and Moldenke, A.R. (1997) Negative effects of ant foraging on spiders in Douglas-fir canopies. Oecologia 109: 313-322.

Hodge, M.A. (1999) The implications of intraguild predation for the role of spiders in biological control. The Journal of Arachnology, 27, 351–362.

Hölldobler B, Wilson E0 (1995) The ants. Springer, Berlin Heidelberg New York

Ives, A.R, Cardinale, B.J., Snyder, W.E. (2005) A synthesis of subdisciplines :predator-prey interactions, and biodiversity and ecosystem functioning. Ecology Letters 8:102-116.

Kajak A, Breymeyer A, Pętal J (1971) Productivity investigation of two types of meadows in the vistula valley. IX Predatory arthropods. Ekol Pol 19: 223-233

Kajak A, Breymeyer A, Pętal J, Olechowicz E (1972) The influence of ants on the meadow invertebrates. Ekol Pol 20: 163-171

Kajak, A., Breymeyer, A., Pêtal, J., & Olechowicz, E. (1972). The influence of ants on the meadow invertebrates. Ekologia Polska, 20 (17), 163-171.

Karhu, K. J. (1998) Effects of ant exclusion during outbreaks of a defoliator and a sap-sucker on birch. Ecological Entomology 23:185-194.

Kling GW, Fry B, O'Brien WJ (1992) Stable isotopes and planktonic trophic structure in Arctic lakes. Ecology 73:561-566

Lajtha K, Michener RH (eds) (1994) Stable isotopes in ecology and environmental science. Blackwell, Oxford

Lang A (2003) Intraguild interference and biocontrol effects of generalist predators in a winter wheat field. Oecologia 134:144-153

Lenoir L, Bengtson J, Persson T (2003) Effects of *Formica* ants on the soil fauna - results from a short-term exclusion and a long-term natural experiment. Oecologia 143:423-430

Lenoir, L. (2003) Response of the foraging behaviour of red wood ants (*Formica rufa* group) to exclusion from trees. Agricultural and Forest Entomology 5:183-189.

Loreau, M., Naeem, S., Inchausti, P., Bengtsson, J., Grime, J.P., Hector, A., Hooper, D.U., Huston, M.A., Raffaelli, D., Schmid, B., Tilman, D. & Wardle, D.A. (2001) Biodiversity and ecosystem functioning: Current knowledge and future challenges. Science, 294, 804–808.

Magnusson, W.E., Carmozina de Araújo, M., Cintra, R., Lima, A.P., Martinelli, L.A., Sanaiotti, T.M., Vasconcelos, H.L., Victoria, R.L. (1999) Contributions of C3 and C4 plants to higher trophic levels in an Amazonian savanna. Oecologia 119:91–96.

Minagawa M, Wada E (1984) Stepwise enrichment of 15-N along food chains: Further evidence and the relation between 15-N and animal age. Geochimi Cosmochimi Acta 48:1135-1140

Naem, S. (2001) in Gardner, R. H., Kemp, W.M., Kennedy, V.S., and Petersen, J. E., editors. (2001) Scaling relations in experimental ecology. Columbia University Press, New York, New York, USA.

Otto D (1965) Der Einfluss der Roten Waldameise (*Formica polyctena* Först.)auf die Zusammensetzung der Insektenfauna (ausschließlich gradierende Arten). Collana Verde 16:250-263

Owens NJP (1987) Natural variations in 15N in the natural environment. Adv Mar Biol 24:389–451

Pętal J, Breymeyer A (1969) Reduction off wandering spiders by ants in a Stellario-Deschampsietum Meadow. Bull Acad Pol Sci Cl. II 17:239-244

Petelle M, Haines B, Haines E (1979) Food preferences analyzed using 13C/12C Ratios. Oecologia 38:159-166

Polis G.A. (1999). Why are parts of the world green? Multiple factors control productivity and the distribution of biomass. Oikos, 86, 3-15.

Polis G.A., & Strong D.R. (1996). Food web complexity and community dynamics. American Naturalist, 147, 813-846.

Polis GA, Myers CA, Holt RD (1989) The ecology and evolution of intraguild predation: potential competitors that eat each other. Annu Rev Ecol Syst 20:297-330

Ponsard S, Arditi R (2000) What can stable isotopes (d15N and d13C) tell about the food web of soil macroinvertebrates. Ecology 81:852-864

Riechert SE, Bishop L (1990) Prey control by an assemblage of generalist predators: Spiders in an Garden test systems. Ecology 71:1441-1450

Riechert, S.E. and Lockeley, T. (1984) Spiders as biological control agents. Ann Rev Entomol 29:299-320.

Riechert, S.E., & Lawrence K. (1997). Test for predation effects of single versus multiple species of generalist predators: spiders and their insect prey. Entomological Experimental Applications, 84, 147-155.

Rosenheim J.A., Kaya, H.K., Ehler, L.E., Marois, J.J., and Jaffee, B.A. (1993) Intraguild predation among biocontrol agents: theory and evidence. Biological Control 5 :303-335.

Rosenheim JA, Wilhoit LR, Armet CA (1993) Influence of intraguild predation among generalist insect predators on the suppression of herbivore population. Oecologia 96:439-449

Schmidt MH, Lauer A, Purtauf T, Thies C, Schaefer M, Tscharntke T (2003) Relative importance of predators and parasitoids for cereal aphid control. Proc R Soc Lond B 270:1905–1909

Schmitz OJ (1998) Direct and indirect effects of predation and predation risk in old-field interaction webs. Am Nat 151:327-342

Schmitz, O.J., Hämback, P.A., & Beckerman, A.P. (2000). Trophic cascades in terrestrial systems: a review of the effects of carnivore removals on plants. American Naturalist, 155, 141–153.

Seifert, B. (2007) Die Ameisen Mittel- und Nordeuropas. Lutra Verlag, Trauer.

Shurin, J.B., Borer, E.T., Seabloom, E.W., Anderson, K., Blanchette, C.A., Broitman, B., Cooper, S.D., & Halpern, B. (2002). A cross-ecosystem comparison of the strength of trophic cascades. Ecology Letters, 5, 785-791.

Slobodkin, L.B., Smith, F.E., Hairston, N.G. (1967) Regulation in terrestrial ecosystems, and the implied balance of nature. American Naturalist 101:109-124.

Snyder, W.E and Ives, A.R (2001) Generalist predators disrupt biological control by a specialist parasitoid. Ecology 82:705-716.

Snyder, W.E., Snyder, G.B., Finke, L.F. & Straub, C.S. (2006) Predator biodiversity strengthens herbivore suppression. Ecology Letters, 9, 789–796.

Symondson, W.O.C., Sunderland K.D., & Greenstone H.M. (2002). Can generalist predators be effective biocontrol agents? Annual Review of Entomology, 47, 561-594.

Tilmann, D, Reich, P.B., Knops, J., Wedin, D., Mielke, T. and Lehmann, C. (2001) Diversity and productivity in a long-term grassland experiment. Science 294:843-845.

Van der Aart P, de Wit T (1971) A field study on interspecific competition between ants (Formicidae) and hunting spiders (Lycosidae, Gnaphosidae, Ctenidae, Pisauridae, Clubionidae) Neth J Zool 21:117-126

Van der Zanden MJ, Rasmussen JB (1999) Primary consumer δ13C and δ15N and the trophic position of aquatic consumers. Ecology 80:1395–1404

Vanderklift MA, Ponsard S (2003) Sources of variation in consumer-diet δ15N enrichment: a meta analysis. Oecologia 136:169-182

Wada E, Mizutani H, Minagawa M (1991) The use of stable isotopes for food web analysis. Crit Rev Food Sci Nutr 30:361-371

Wilby, A. and Thomas, M.B. (2002) Natural enemy diversity and pest control: patterns of pest emergence with agricultural intensification. Ecology Letters 5: 353–360.

Wise, D.H. (1993) Spiders in ecological webs. Cambridge University Press, Cambridge.

Chapter 2

Intraguild interactions between spiders and ants and top-down control in a grassland food web

Dirk Sanders and Christian Platner

Oecologia 2007, 150, 611–624

Abstract

In most terrestrial ecosystems ants (Formicidae) as eusocial insects and spiders (Araneida) as solitary trappers and hunters are key predators. To study the role of predation by these generalist predators in a dry grassland, we manipulated densities of ants and spiders (natural and low density) in a two-factorial field experiment using fenced plots. The experiment revealed strong intraguild interactions between ants and spiders. Higher densities of ants negatively affected the abundance and biomass of web-building spiders. Density of Linyphiidae was three times higher in plots without ant colonies. The abundance of *Formica cunicularia* workers was significantly higher in spider-removal plots. Also, population size of springtails (Collembola) was negatively affected by the presence of wandering spiders. Ants reduced the density of Lepidoptera larvae. In contrast, the abundance of coccids (Ortheziidae) was positively correlated with densities of ants.

To gain a better understanding of the position of spiders, ants and other dominant invertebrate groups in the studied food web and important trophic links, we used a stable isotope analysis (^{15}N and ^{13}C). Adult wandering spiders were more enriched in ^{15}N relative to ^{14}N than juveniles, indicating a shift to predatory prey groups. Juvenile wandering and web-building spiders showed $δ^{15}$N ratios just one trophic level above those of Collembola and had similar $δ^{13}$C values, indicating that Collembola are an important prey group for ground living spiders. The effects of spiders demonstrated in the field experiment support this result.

We conclude that the food resource of spiders in our study system is largely based on the detrital food web and that their effects on herbivores are weak. The effects of ants are not clear-cut and include predation as well as mutualism with herbivores. Within this diverse predator guild intraguild interactions are important structuring forces.

Keywords Field experiment, Generalist predators, Stable isotopes, Collembola, $δ^{15}$N /$δ^{13}$C

Introduction

In terrestrial ecosystems, spiders and ants are ubiquitous and diverse generalist predators (Wise 1993; Hölldobler and Wilson 1995). Most Central European ant species are omnivores, being able to prey on a wide range of other invertebrates, as well as to take up nutrients from plants indirectly by trophobiosis with phloem-feeding insects (Seifert 1996). Spiders and ants are potential competitors and mutual predators. Intraguild predation, i.e. feeding on species of the same guild, is common in natural communities (Polis et al. 1989) and enhances the reticulate nature of a food web. Further, intraguild predation has been characterised as an important feature structuring arthropod communities (Wise 1993).

Studies have reported high rates of predation by ants on spiders (Pętal and Breymeyer 1969; Kajak et al. 1972), but there is a lack of evidence for these effects demonstrated by field experiments (Wise 1993). Halaj et al. (1997) tested the effect of ants foraging on a spider assemblage in Douglas-fir canopies. The abundance of hunting spiders increased significantly following ant exclusion. However, the authors concluded that not direct predation but disturbance of spiders by ants was important. In contrast, such effects could not be demonstrated in other studies (Otto 1965; van der Aart and de Wit 1971; Brüning 1991; Lenoir et al. 2003; Gibb 2003). Brüning (1991) tested the effects of *Formica polyctena* on spiders in a forest ecosystem without finding any difference in density or composition of the spider community neither inside nor outside the hunting area of ants.

In the current study, we manipulated densities of spiders and ants in a field experiment and tested their effects as predators in a diverse arthropod community. Additionally, we used a stable isotope analysis to gain a better understanding of trophic links in the food web. Stable isotope analysis of ratios of $^{15}N/^{14}N$ and $^{13}C/^{12}C$ is a promising tool for food web studies (De Niro and Epstein 1981;Wada et al. 1991; Kling et al 1992; Ponsard and Arditi 2000). Values of $\delta\ ^{13}C$ are largely conserved in food chains and provide information about the identity of the resource base (DeNiro and Epstein 1978; Petelle et al. 1979; Magnusson et al. 1999; Vander Zanden and Rasmussen 1999), whereas $\delta^{15}N$ values can be used as a trophic level indicator (Ponsard and Arditi 2000; Post 2002; Vanderklift and Ponsard 2003).

Materials and methods

Study site

The experiment was conducted on a limestone hillside (51°22´N, 9°50´E) near Witzenhausen (Hesse, Germany) exposed to the south. The long-term mean temperature in January is 0°C and 18°C in July and the annual precipitation amounts to approximately 650 mm (Stein 1996). The experimental area comprised a dry grassland (Mesobromion) and a meadow (Arrhenatheretum) and had not been in use as pastureland throughout the last ten years (for details of vegetation see appendix 1). The experimental area was located 180 – 200 m above sea level adjacent to a mixed beech and pine forest and was surrounded by bushes. The density and height of the herb layer increased downhill from the area of the dry grassland to the meadow.

We found 72 spider species, with wolf spiders and web-building spiders such as linyphiids and araneids being most abundant. Among the 18 species of ants in the study site, the most abundant were *Myrmica sabuleti* Meinert, *Lasius alienus* Förster and the subterranean species *Lasius flavus* (Fabricius), with medium worker densities of all species combined outside the stricter nest areas of 500 – 700 individuals/m^2. Nest distribution of less abundant species was very patchy. The herbivorous guild in the grassland was a diverse mixture of species consisting mainly of grasshoppers, planthoppers, leafhoppers, beetles, heteropteran bugs and aphids.

Experiment

The basic experimental unit was a 2 m^2 area, enclosed by a 30-cm high plastic fence. The fence surrounding these plots was buried 10 cm deep into the ground and equipped with slippery barriers of silicon gel on the inner side of the fence to reduce emigration from non-removal-plots and on the outside to prohibit immigration of spiders and ants in removal plots (Oraze & Grigarick 1989). The experiment ran from May until September 2002 and was set up in a two-factorial design with two levels of spider and ant density (natural and low), resulting in four treatment combinations. Each combination was replicated five times in blocks giving a total of 20 plots. The five blocks formed a transect from the top to the bottom of the hillside, each being located in different vegetation in the gradient of the dry grassland down to the meadow (see appendix 1).

The low predator-density treatment was achieved indirectly by placing slippery barriers on the outside of the rings and by removing spiders manually and excluding ant colonies. Spider populations and ant colonies that became re-established in the removal plots were removed twice a week during the four months of the experiment. One person searched each plot for spiders and ant colonies for ten minutes. Detected ant colonies in these plots were excavated and replaced by soil cores without ants from outside the plots. In ant plots with only one colony a supplementary colony of *Lasius* or *Myrmica* that was excavated outside the plots was added to achieve a comparable ant nest density. On average, 3 to 6 spiders per plot were removed from low spider-density treatments on each sampling occasion and released to the remaining non-removal-plots in the same block. To assess the effect of enclosures, for each of the five blocks one sample was taken outside the plots in similar vegetation. A comparison with the control samples suggested that spider densities and biomass and ant biomass reached a natural level in non-removal-plots (see Figures 1 and 2a). Both wandering and web-building spiders were removed, but we achieved no reduction of web-building spider density in removal plots (see results).

Sampling

The fauna was sampled in June, August and September 2002 with a suction sampler (Stihl SH 85, Germany; 10 s/sample using a 0.036 m² sampling cylinder) and additionally on two occasions (June and September) by heat extraction from 0.036 m² soil cores (Kempson 1963, Schauermann

1982). One sample per plot was taken on each occasion. Spiders, ants, planthoppers and leafhoppers were identified to species level, while other arthropods were assigned to higher-ranking taxa. Spiders were separated into two functional groups: web-building spiders and wandering spiders. All spiders and ants found in the samples were dried for 72 h at a temperature of 60°C, and the dry weight of ants and spiders was measured. On two occasions, in June and August, the number of spider webs in the plots was counted to assess the activity of web-building spiders.

Data analyses

The effects of the spider and ant treatment and the response of the diverse arthropod community were analysed by a repeated measures two factor analysis of variance (rmANOVA) (Ende 1993). For large-sized Collembola and for Lepidoptera larvae with data for only one sampling occasion we performed a two factor ANOVA. For ants the sum of all soil and suction samples was analysed because suction samples on their own were insufficient to record the abundance of ground-living ants. All abundance and biomass data were log-transformed to meet assumptions of normality and homogeneity of variances.

Stable isotopes

Ratios of 13C and 15N were estimated by a coupled system consisting of an elemental analyzer (Carlo Erba NA 2500) and a gas isotope mass spectrometer (Finnigan Deltaplus). The system is computer-controlled allowing measurement of ^{13}C and ^{15}N (Reineking et al. 1993). Isotopic contents were expressed in δ units as the relative difference between sample and conventional standards with $δ^{15}$N or $δ^{13}$C [‰] = (RSample − RStandard)/RStandard x 1000, where R is the ratio of ^{15}N/^{14}N or ^{13}C/^{12}C content, respectively. The conventional standard for ^{15}N is atmospheric nitrogen and for ^{13}C PD-belemnite (PDB) carbonate (Ponsard and Arditi 2000). Acetanilide (C8H9NO, Merck, Darmstadt) served for internal calibration with a mean standard deviation of samples <0.1‰. Dried samples were weighed into tin capsules to contain 500-1800 μg of dry biomass and stored in a desiccator until measurement. For the large spider genera *Alopecosa*, *Pisaura* and *Atypus* it was necessary to use only parts of the body (prosoma), while small individuals of juvenile spiders and springtails were combined into one sample. Albers (2002) analyzed parts of the body of arthropods and found no significant differences in their $δ^{15}$N values. If possible, replicate measurements were made. We analysed spiders and ants, their potential prey and plants. Plants from the soil cores were separated into herbs, grasses and mosses, and samples of these groups were replicated six times. Stable isotope data were analysed by performing a general linear model (GLM) due to different size of samples. All statistical analyses were performed with SAS (Version 8: proc glm and proc anova). *Aulonia albimana* (Lycosidae), which was one of the most abundant spiders and present in all samples, was used for the comparison of possible block- and treatment-specific differences in stable isotope ratios. No such differences between the five blocks and treatments, including non-fenced controls, were found

($\delta^{13}C$: for treatment $F_{4,16}$ = 1.35, P = 0.29 and block $F_{4,13}$ = 0.93, P = 0.48; $\delta^{15}N$: for treatment $F_{4,16}$ = 0.59, P = 0.68 and block $F_{4,13}$ = 1.72, P = 0.20; GLM).

Results

Manipulation of spider and ant density

During the experiment 964 spiders were captured and removed in the spider-removal plots (about 700 wandering spiders, 260 web-building spiders). There was a significant effect of spider removal on the total abundance and biomass of wandering spiders (Fig. 1a; Table 1). Biomass and density of wandering spiders was 2.4 times lower in spider removal-plots. The effect on the biomass tended to be more pronounced in June than in August and September (Fig. 1a; Table 1). In contrast, biomass and density of web-building spiders were not affected by the manipulation (Fig. 1b; Table 1).

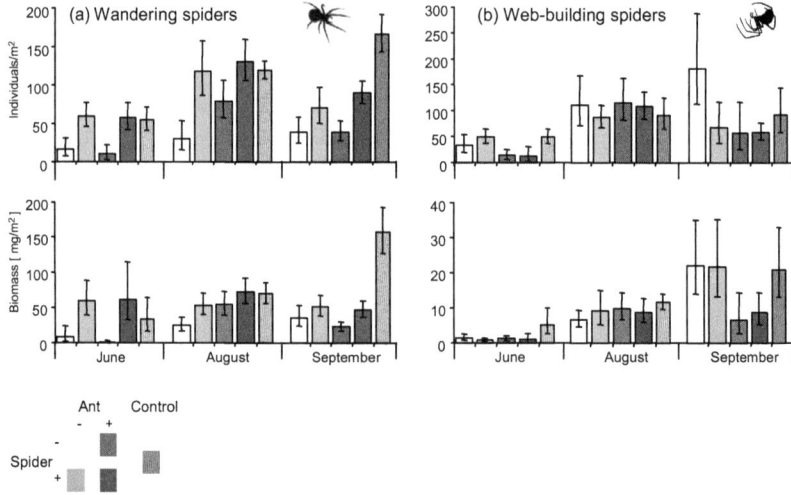

Fig. 1 Mean abundance and biomass of wandering spiders (a) and web-building spiders (b) in suction-samples from the four different treatment combinations with natural and reduced ant and spider density and in control samples outside the plots. Open bars: plots with reduced spiders and ant density; shaded bars: plots with natural spider density and hatched bars refer to plots with natural ant density; dotted bars: controls from outside the plots. Geometrical means (n=5); *error bars are back-transformed standard errors of the mean ignoring the block effect*. For statistical analyses see text and table 1.

The total biomass of all epigeic active ants was successfully manipulated (Fig. 2a; Table 2). The biomass of ants and wandering spiders in non-removal plots was not significantly different from biomass values in control samples outside the plots (ant biomass: $F_{1,8}$ = 0.13, P = 0.72 for the effect of treatment in a one factor ANOVA; spider biomass: $F_{1,8}$ = 0.17, P = 0.69 for the effect of treatment in a rmANOVA).

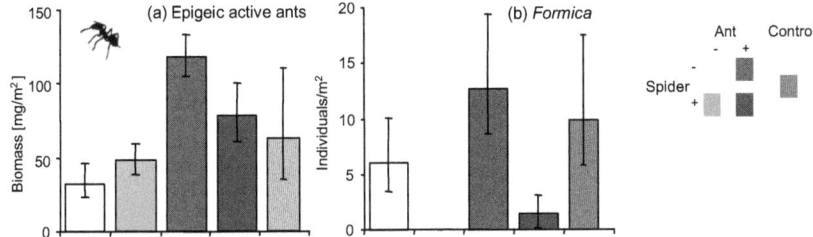

Fig. 2 (a) Biomass of epigeic active ants (geometrical means of total biomass of all epigeic ant species sampled per plot) and (b) abundance of Formica cunicularia and F. fusca (suction samples from August) in the four different treatment combinations with natural and reduced ant and spider density and in control samples outside the plots. Open bars: plots with reduced spiders and ant density; shaded bars: plots with natural spider density and hatched bars refer to plots with natural ant density; dotted bars: controls from outside the plots. Geometrical means (n = 5); *error bars are back-transformed standard errors of the mean ignoring the block effect.* For statistical analyses see table 2.

Table 1 Response of wandering and web-building spiders to the treatments. Data were log-transformed (log10X+1). F values are given for a repeated measures ANOVA for suction samples from June, August and September; for the within effects F values for Pillai's Trace are given. df = degrees of freedom (Nom, Den), bold digits indicates statistical significance (p<0.05).

	df	Wandering spiders (abundance)		Wandering spiders (biomass)		Web-building spiders (abundance)		Web-building spiders (biomass)	
		F	P	F	P	F	P	F	P
Ant (A)	1, 12	0.14	0.7180	0.21	0.6582	3.47	0.0873	0.91	0.3581
Spider (S)	1, 12	10.90	**0.0063**	17.46	**0.0013**	0.48	0.5022	0.00	0.9816
A × S	1, 12	0.01	0.9444	0.54	0.4760	0.26	0.6218	0.00	0.9597
Block (Bl)	4, 12	0.57	0.6866	1.99	0.1604	2.83	0.0730	2.89	0.0690
Time (T)	2, 11	6.79	**0.0120**	4.82	**0.0314**	9.27	**0.0044**	35.48	**<0.0001**
T × A	2, 11	1.08	0.3727	2.69	0.1117	3.31	0.0750	3.88	0.0531
T × S	2, 11	0.65	0.5392	4.14	**0.0458**	0.66	0.5368	0.31	0.7419
T × A × S	2, 11	1.18	0.3425	1.18	0.3424	0.99	0.4012	0.35	0.7103
T × Bl	8, 24	1.09	0.4038	1.18	0.3517	1.85	0.1165	1.07	0.4132

Interactions between ants and spiders

The presence of ant colonies had a negative impact on web-building spider abundance in June (Fig. 1b; $F_{1,12}$ = 7.72, P = 0.017; for ANOVA). Biomass of web-building spiders in suction-samples showed a negative response to higher ant densities only in September ($F_{1,12}$ = 5.50, P = 0.037; for rmANOVA incl. interaction ant x time, see Table 1). Samples taken by heat extraction from soil cores revealed a negative effect of ants on web-building spiders, most of which belonged to the Linyphiidae (Fig. 3b, Table 3). Higher ant densities also significantly decreased the number of established linyphiid webs (Fig. 3a, Table 3). In June the density of web-building spiders was 2.5 times and in September 3 times higher in ant-removal plots (samples taken by suction trap, Fig. 1b). The mean number of web-building spiders in the ant and spider-removal plots in September was 180 individuals per m^2 compared to 60 individuals per m^2 in non-removal plots of spiders and ants, but these effects of ants and wandering spiders on the abundance of web-building spiders were not significant (Table1). Abundance and biomass of wandering spiders were not affected by the presence of ants (Fig. 1a, Table 1).

In August, the abundance of *Formica cunicularia* Latreille and *F. fusca* L. workers reached higher densities in spider-removal plots (Fig. 2b; Table 2). *Formica* colonies were not present inside the plots; abundance ranged from 7 to 13 individuals of *Formica* per m^2 in plots excluding wandering spiders.

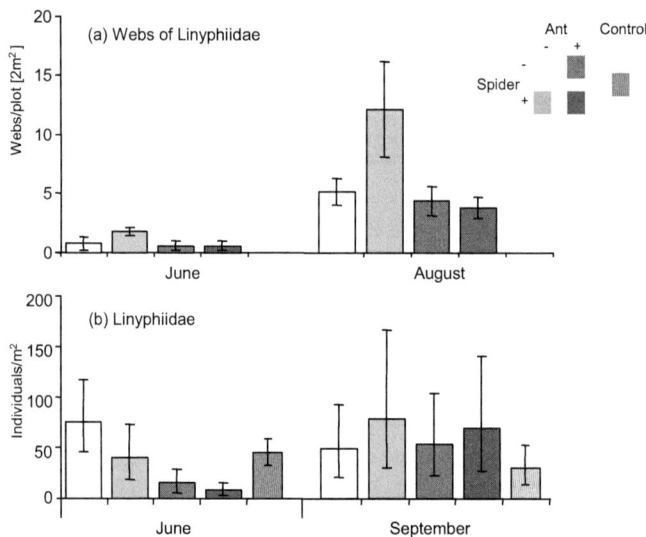

Fig. 3 Mean numbers of (a) spider webs and (b) linyphiids (samples from soil cores) in the four different treatment combinations with natural and reduced ant and spider density and in control samples outside the plots. Open bars: plots with reduced spiders and ant density; shaded bars: plots with natural spider density and hatched bars refer to plots with natural ant density; dotted bars: controls from outside the plots. For webs (a) arithmetic means are given (n = 5) and error bars are standard errors of the mean, for Linyphiidae (b) geometrical means are given (n = 5) and error bars are back-transformed standard errors of the mean, both ignoring the block effect. For statistical analyses see table 3.

Effects of ants and spiders on the arthropod community

Higher densities of wandering spiders had a negative effect on the density of epigeic Collembola. In June, wandering spiders negatively affected the abundance of Collembola species larger than 1 mm (Fig. 4a; Table 2) but not the abundance of all Collembola (Fig. 4b; Table 3). The density of Collembola increased by 37% in ant-removal plots compared to that in natural-ant density plots in June, but the difference was only marginally significant ($F_{1,12}$ = 3.97, P = 0.0696), whereas the abundance of larvae of Lepidoptera responded negatively to the presence of ants in September (Fig. 4c; Table 3). For planthoppers and leafhoppers effects of spiders were not found (Fig. 4d; Table 3). We also could not find any effect on the total number of Isopoda, Julidae, Geophilidae, Lithobiidae, Heteroptera, aphids, beetles, dipterans.

The phloem-feeding Ortheziidae (coccids) showed a positive response to higher densities of ants (Fig. 4e; Table 3). The abundance of Ortheziidae increased significantly from June to September (Table 3). Thysanoptera (thrips) showed a similar negative response to ant removal (Fig. 4f), however, the response was only significant in September ($F_{1,12} = 6.90$, $P = 0.0221$; for ANOVA). This effect was dependent on spider treatment and time (significant time x ant x spider interactions; Table 3).

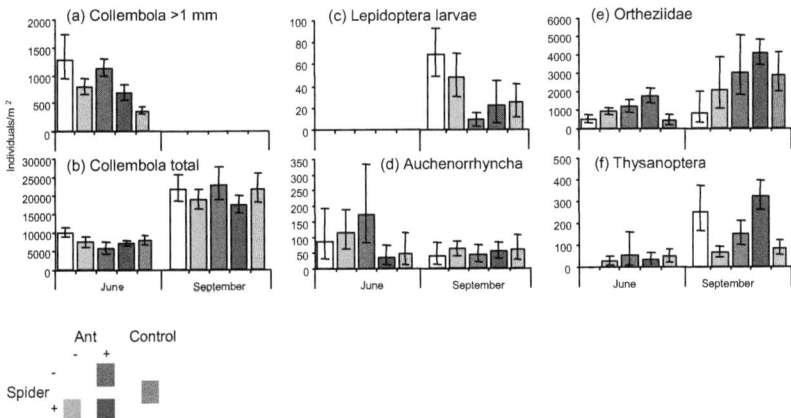

Fig. 4 Mean abundance of Collembola larger than 1 mm (a), all Collembola (b), Lepidoptera larvae (c), Auchenorrhyncha (d), Ortheziidae (e) and Thysanoptera (f) in samples from soil cores taken in the four different treatment combinations with natural and reduced ant and spider density and in control samples outside the plots. Open bars: plots with reduced spiders and ant density; shaded bars: plots with natural spider density and hatched bars refer to plots with natural ant density; dotted bars: controls from outside the plots. Geometrical means (n=5); error bars are back-transformed standard errors of the mean ignoring the block effect. For statistical analyses see tables 2 and 3.

Table 2 Response of ants (sum of all epigeic species and abundance of *Formica cunicularia* and *F. fusca* workers from suction-samples), Collembola and Lepidoptera larvae (from soil samples), using a two-way ANOVA. Data were log-transformed (log10X+1). df = degrees of freedom, bold digits indicates statistical significance (p<0.05).

	df	Epigeic ants (biomass; sum of samples)		*Formica* (worker-abundance; August)		Collembola >1mm (abundance; June)		Lepidoptera larvae (abundance; September)	
		F	P	F	P	F	P	F	P
Model	7, 12	4.83	**0.0085**	3.23	**0.0361**	1.53	0.2457	1.78	0.1818
Ant	1, 12	19.03	**0.0009**	2.28	0.1569	0.44	0.5174	6.84	**0.0226**
Spider	1, 12	0.003	0.9607	13.60	**0.0031**	5.58	**0.0359**	0.01	0.9109
A × S	1, 12	3.87	0.0727	0.45	0.5170	0.00	0.9691	1.06	0.3228
Block	4, 12	2.73	0.0792	1.57	0.2448	1.18	0.3692	1.13	0.3866

Table 3 Response of members of the arthropod community from soil and litter layer (heat extraction of soil cores) to the biomass manipulation of ants and wandering spiders. All comparisons were made using a two-way repeated measures ANOVA for data from June and September; for the within effects F values for Pillai's Trace are given. Data were log-transformed (log10X+1). df = degrees of freedom (Nom, Den), bold digits indicates statistical significance (p<0.05).

	df	Linyphiidae (abundance)		Linyph. webs		Collembola (total)		Auchenorrhyncha		Ortheziidae		Thysan.	
		F	p	F	p	F	p	F	p	F	p	F	p
Ant (A)	1, 12	7.49	**0.0180**	5.63	**0.0352**	1.90	0.1935	0.02	0.8969	5.20	**0.0416**	4.11	0.0655
Spider (S)	1, 12	0.03	0.8593	2.74	0.1235	1.04	0.3287	0.02	0.8969	2.17	0.1665	0.01	0.9310
A × S	1, 12	0.01	0.9234	3.71	0.0783	0.85	0.3753	0.66	0.4310	0.32	0.5823	0.51	0.4886
Block (Bl)	4, 12	6.31	**0.0057**	1.40	0.2920	2.28	0.1207	1.64	0.2274	1.62	0.2326	2.77	0.0770
Time (T)	2, 11	2.44	0.1446	26.72	**0.0002**	137.0	**<0.0001**	7.24	**0.0196**	13.10	**0.0035**	46.7	**<0.0001**
T × A	2, 11	1.70	0.2167	3.42	0.0892	3.02	0.1078	0.46	0.5104	0.34	0.5700	0.19	0.6717
T × S	2, 11	0.90	0.3612	1.64	0.2246	0.91	0.3595	0.94	0.3505	0.04	0.8412	0.83	0.3813
T × A × S	2, 11	0.16	0.6953	2.45	0.1436	3.77	0.0760	1.19	0.2968	0.20	0.6655	13.1	**0.0035**
T × Bl	8, 24	0.80	0.5452	2.49	0.0993	3.00	0.0622	4.23	**0.0230**	3.15	0.0549	4.29	**0.0220**

Analysis of stable isotopes

The plant groups had $\delta^{13}C$ values of –28.5 to –30‰ and $\delta^{15}N$ values of –5 to –3‰ (Fig. 5). Most herbivorous insects such as planthoppers, leafhoppers, Ortheziidae and aphids showed $\delta^{15}N$ values very similar to plants. The Alticinae (Chrysomelidae) were more enriched in ^{15}N with a $\delta^{15}N$ value of –2.2 ‰. Detritivorous and fungi-feeding arthropods such as Julidae, Isopoda and Collembola with a $\delta^{15}N$ value of –1.84‰, had higher $\delta^{13}C$ values than plants.

Wandering spiders, consisting of *Aulonia albimana* (Walckenaer) and juvenile spiders of the genus *Zora*, *Tibellus*, *Pardosa* and *Clubiona*, were 2 - 3 ‰ more enriched in ^{15}N than Collembola ($F_{1,53}$ = 55.78, P < 0.001, for GLM). Both groups had similar $\delta^{13}C$ values ($F_{1,53}$ = 0.32, P = 0.57, for GLM; Fig. 5a). Juvenile web-building spiders and adult *Tenuiphantes tenuis* (Blackwell) (Fig. 5b) were also more enriched in $\delta^{15}N$ than Collembola ($F_{1,13}$ = 19.88, P < 0.001, for GLM) and had similar $\delta^{13}C$ values ($F_{1,13}$ = 0.21, P = 0.66, for GLM). *Walckenaeria acuminata* Blackwell, *Atypus piceus* (Sulzer), *Alopecosa trabalis* (Clerck), *Pisaura mirabilis* (Clerck) and *Tibellus oblongus* (Walckenaer) were more enriched in $\delta^{15}N$ with values 4 to 5 ‰ higher than Collembola. Among all spiders *Atypus* was most enriched in $\delta^{13}C$. The spiders most enriched in ^{15}N were the web-building species *Argiope bruennichi* (Scopuli), *Mangora acalypha* (Walckenaer) and the wolf spiders *Arctosa lutetiana* (Simon) and *Pardosa lugubris* (Walckenaer) with $\delta^{15}N$ values higher than 4. Generally $^{15}N/^{14}N$ ratios in adult wandering spiders (*Pisaura*, *Pardosa*, *Tibellus*) were significantly higher than in juveniles (Fig. 5a, $F_{1,9}$ = 8.63, P = 0.016, for GLM).

Among ant species, *Lasius flavus* and *L. alienus* had lower $^{15}N/^{14}N$ ratios than *Myrmica sabuleti*, *Formica cunicularia* and *Ponera coarctata* (Latreille) (Fig. 5c). $\delta^{13}C$ values of the *Lasius* species had a higher variance in comparison to *Formica* and *Myrmica*. Among the generalist predators adult wolf spiders and *Atypus* contained higher values of $\delta^{15}N$ (Fig. 5d) than most web-building spiders and ants ($F_{1,79}$ = 46.32, P < 0.001, for GLM). Values of all arthropods analysed can be found in the appendix 2. Diptera of the family Sphaeroceridae were most enriched in ^{15}N with $\delta^{15}N$ values of 5.74.

Intraguild interactions between spiders and ants

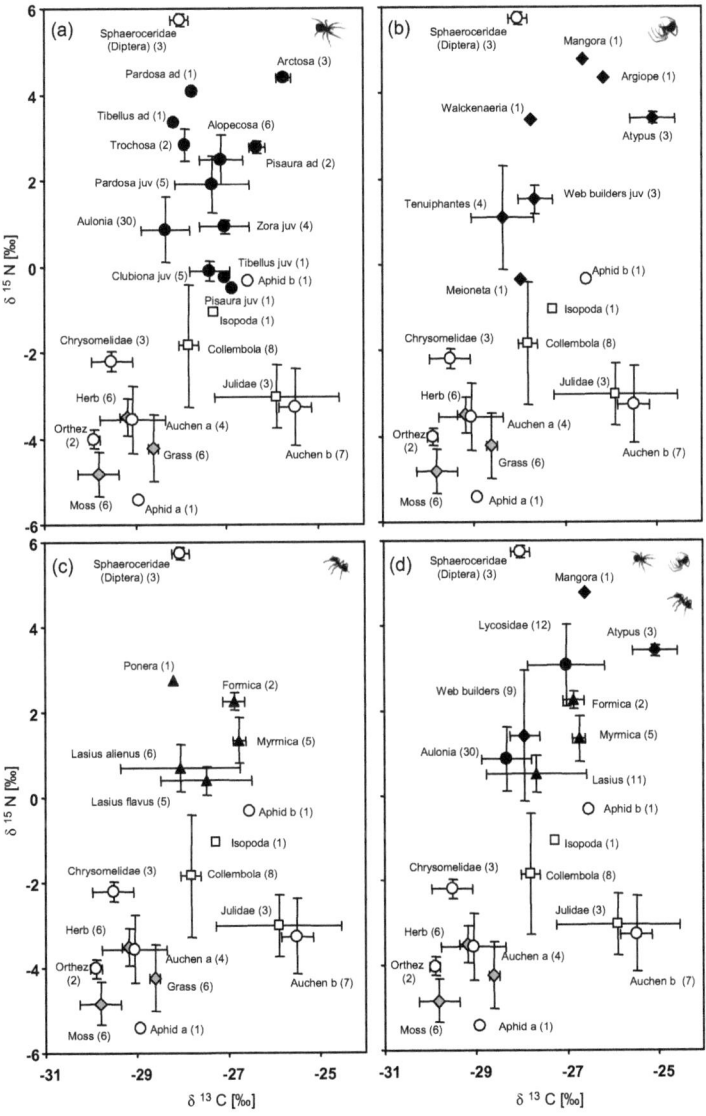

Fig. 5 δ^{15}N and δ^{13}C values (± SD) of wandering spiders (a), web-building spiders (b), ants (c), of their possible prey organisms (open circles) and of plants (shaded diamond). Numbers of samples analysed are given in parentheses. A summary for the most important predatory groups is presented in (d). Web builders juv = juvenile web-building spiders (Linyphiidae, Theridiidae, Tetragnathidae, Araneidae), Auchen = Auchenorrhyncha. Full names and values of the taxa referred to in this figure are given in appendix 2. ● = wandering spiders, ♦ = web-building spiders, ▲ = ants, ○ = herbivores, □ = detritivorous and fungivorous groups.

Discussion

Manipulation of spider and ant densities

We successfully manipulated densities and biomass of wandering spiders and ants. Natural spider density treatment was achieved by spider addition to the non-removal plots because the enclosures seemed to have had a negative effect on spider populations. Wandering spider density and biomass in non-removal plots was similar to control samples in June and August, but tended to be lower in September. We assume that the barriers of the fence and silicon gel did not prevent all spiders from leaving the plots. An alternative explanation is an enhancement of cannibalism at higher spider densities. Apparently, in most cases it seems to be impossible to achieve densities higher than the natural densities of spiders in a long-term experiment (Wise 1993). We could not reduce density and biomass of web-building spiders in removal-plots, probably due to an increased survival in plots without wandering spiders.

Intraguild interactions

Intraguild predation has been identified as an important feature structuring terrestrial arthropod communities, in particular if spiders are involved (Wise 1993). However, most studies did not provide evidence of interactions between spiders and ants affecting population densities (Otto 1965; van der Aart and de Wit 1971; Brüning 1991; Lenoir et al. 2003; Gibb 2003).

Our results provide experimental evidence for negative interactions between ants and spiders in a grassland (Fig. 6). We observed a negative effect of ants on the abundance of web-building spiders. This effect was strong in June and September, with densities of web-building spiders being up to 3 times higher in ant-removal plots. This coincides with periods of high predation by *Myrmica* in the time of intensive growth of ant larvae, as reported by Kajak et al. (1971). The density of spider webs in the herb layer was also significantly lower in plots with ant colonies, indicating lower activity of web-building spiders. Lenoir et al. (2003) found a similar negative effect of *Formica rufa* on the activity of Linyphiidae on the forest floor after excluding ants from their usual food sources in the tree canopy and thus forcing them to forage on the ground. In our experiment the effects of ants on web-building spiders were stronger in samples from soil cores in comparison to suction samples, indicating that ants had a greater impact on ground dwelling spiders than on those in higher strata of the herb layer. The majority of web-building spiders were sheet-web weavers of the subfamilies Linyphiinae and Erigoninae. These spiders build their cryptic webs in the litter layer as juveniles and live within easy reach of foraging ants.

In August, higher densities of wandering spiders led to a decrease in the abundance of the ants *Formica cunicularia* and *F. fusca*. This effect could be substantiated only for ants away from their colonies, since no colonies were present within our plots. Both *Formica* species seem to be less aggressive than *Myrmica* spp. and *Lasius alienus* (Seifert 1996); probably foraging workers of *Formica* species avoid areas of higher densities of wandering spiders due to a higher disturbance rate. Such trait mediated effects caused by disturbance seem to be important in arthropod communities as recently demonstrated for spiders and other prey groups, e.g. grasshoppers, planthoppers and leafhoppers (Schmitz 1998; Cronin et al. 2004). However, Brüning (1991) observed some species of Theridiidae, Amaurobiidae and Segestriidae preying upon workers of *Formica*. In our study, feeding activity of the spiders was not directly assessed, but we observed a few individuals of Lycosidae and Thomisidae preying upon ants.

We conclude that intraguild interactions were important forces for structuring the community. Web-building spiders reached highest densities in ant-removal plots, indicating a negative influence of ants (Fig. 6). Further, ants of the genus *Formica* were negatively affected by the presence of wandering spiders.

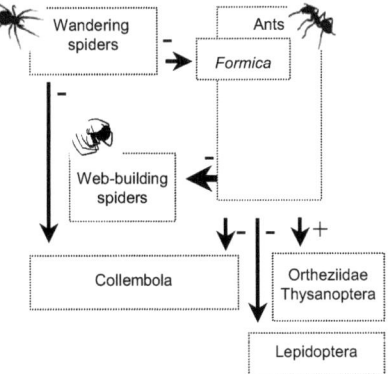

Fig. 6 A model of important interactions between arthropod groups in the grassland food web. Effects were tested by ANOVA, for strong effects p< 0,05.

Top-down control

There is a growing body of evidence for an important role of ants and spiders as controlling forces for other grassland arthropods (Kajak et al. 1972; Wise 1993; Riechert and Lawrence 1997). In their study of the role of *Myrmica* in a meadow ecosystem, Kajak et al. (1972) reported high predation rates of ants on juvenile arthropods. In our study we observed a negative effect of ants on larvae of Lepidoptera and on Collembola but a positive effect on Ortheziidae and Thysanoptera (Fig. 6). In contrast, for wandering spiders we observed only an effect on the abundance of large-sized Collembola but not on any group of herbivorous arthropods. Lawrence and Wise (2000; 2004) and Wise (2004) demonstrated that experimental removal of wandering spiders in the field significantly increased the abundance of Collembola. Wolf spiders consume Collembola in an amount ranging between 8% and 40% of total spider's diet (reviews in Nentwig 1986 and Nyffeler 1999). In our study, $\delta^{13}C$ values of Collembola and many ground living spider species (wandering spiders and juvenile web builders) were similar, suggesting that these spiders feed on Collembola to a significant extent. Juvenile wandering and web-building spiders in our study showed $^{15}N/^{14}N$ ratios just one trophic level above those of Collembola, indicating that Collembola are an important prey group.

A comparison of our results with studies of agro-ecosystems (Sigsgaard 2002; Agusti et al. 2003) suggest that generally the food resource of ground-living spiders is based mainly on the components of the detrital food web (notably Collembola and other detritivores) and that effects on herbivore populations are often weak. In contrast other studies found strong effects of spiders on pest species in agro-ecosystems (Riechert and Bishop 1990; Snyder and Wise 2001; Schmidt et al. 2003). But in these studies the guild of herbivores only consisted of few pest species, whereas in our study the herbivore guild as well as the predator guild was very diverse mixtures of many species from different taxa.

Attention must also be paid to the indirect effects of intraguild predation, which have recently been shown to reduce the strength of top down effects (Lang 2003; Finke and Denno 2003; 2004; Denno et al. 2004), and therefore it may not be possible to demonstrate top down effects on the herbivorous guild in highly diverse systems. Effects on herbivores and spiders in an other recent ant exclusion experiment were not strong, probably due to a compensatory change in the composition of the invertebrate predator guild (Laakso and Setälä 2000). The removal of ants and wandering spiders in our experiment caused high web-building spider densities. The nearly constant presence of generalist predators in all treatments may result in a constant overall top down control by these three different predator groups. This effect of a highly diverse predator guild may stabilize the whole system, as indicated by the weak overall effects of the clearly reduction of single predator groups.

Food web analysis

On average, the $^{15}N/^{14}N$ ratio of predators is increased by 3–4‰ compared with their prey (DeNiro and Epstein 1981; Minagawa and Wada 1984; Owens 1987; Peterson and Fry 1987; Cabana and Rasmussen 1994). However, within this general pattern variation in consumer diet $\delta^{15}N$ enrichment can be substantial (Vanderklift and Ponsard 2003). Our data indicate that the food web in the dry grassland may span three trophic levels (values between –5.4 and +5.7‰ $\delta^{15}N$). In spiders values ranged from –0.5 to +4.8‰ $\delta^{15}N$, with a high overlap and variance of $^{15}N/^{14}N$ ratios. Hence we could not assign spiders to a single trophic level. Adults were more enriched in ^{15}N than juveniles, indicating different trophic positions in the food web. In contrast to $^{15}N/^{14}N$, $^{13}C/^{12}C$ ratios of plants tend to pass along the food chain with little further fractionation and are only slightly enriched in higher trophic levels (DeNiro and Epstein 1978, Petelle et al. 1979, Macko et al. 1982; Minagawa and Wada 1984; Lajtha and Michener 1994). The content of ^{13}C in the tissue of predators resembles that of their food (DeNiro and Epstein 1978) and can be used to identify the food resource (Magnusson et al. 1999; Vander Zanden and Rasmussen 1999). Apparently, in particular juvenile spiders largely prey upon Collembola as can be inferred from their very similar $\delta^{13}C$ values. A similarly tight trophic connection between small spiders and Collembola was also found by McNabb et al. (2001) in an agro-ecosystem. In contrast, in adult individuals of *Pisaura* and wolf spiders $\delta^{15}N$ values more than one trophic level above Collembola indicate that they often feed on predatory arthropods, which probably include other spiders and members of their own species. As predators grow, the size range of utilised prey may change, and may include smaller individuals of other predatory species (Rosenheim et al. 1993). However, Oelbermann and Scheu (2002) found a significantly lower ^{15}N content of hatchlings of the wolf spiders *Pardosa lugubris* than that of their mothers, indicating the existence of nitrogen pools with different ^{15}N signatures in female *P. lugubris*. Among all spiders *Atypus piceus* showed the highest δ^{13}C values, and may therefore be more strongly connected to the soil food web. This species builds a silken tube reaching from belowground to the soil surface and is supposed to prey upon arthropods such as Julidae and Isopoda, which move over the tube. This interpretation is supported by the higher δ^{13}C values of these two prey groups in comparison to other possible prey groups.

Ants that are mainly predatory, such as *Formica cunicularia* and *Ponera coarctata* were more enriched in ^{15}N. In contrast, *Lasius flavus* and *L. alienus* had a lower ^{15}N/^{14}N ratio, probably due to higher rates of trophobiosis with aphids or coccids. A similar relationship was demonstrated for a rainforest ant community (Blüthgen et al. 2003). The authors found that δ^{15}N values for ant species that commonly forage for nectar were low while predominantly predatory species showed high values. The positive effects of higher ant densities, especially of *Lasius flavus*, on the abundance of Ortheziidae are in accordance with known interactions between this ant species and other aphids on plant roots groups (Seifert 1996). Among all studied arthropods, Diptera of the family Sphaeroceridae were the most enriched in ^{15}N. Larvae of most species are known to feed on dung or other decaying matter of plants and animals (Pitkin 1988; Smith 1989). As a consequence, high δ^{15}N values may result from dead animal material in their diet.

For juvenile wandering spiders and for ground living web-building spiders, Collembola were a key resource. This finding is supported by top down effects revealed by the field experiment and the stable isotope analysis. Additionally, we found top-down effects of ants on Lepidoptera larvae and on Collembola. However, the effects of ants included predation as well as mutualism with sap-feeding herbivores. The food resource of most generalist predators in our study system is largely based on the detrital food web, at least temporarily.

Acknowledgments

We thank David H. Wise (Kentucky), Matthias Schaefer, Herbert Nickel, Sonja Migge (all Göttingen), Eike Gentsch (Bremerhaven) and two anonymous referees for valuable discussions and comments on the manuscript. We are grateful to Sharon Cooper, Terence Kleian, Tristan Ernsting (Göttingen) for linguistic corrections. Gerald Moritz (Halle) and Klaus Hövemeyer (Göttingen) provided valuable comments on the biology of thrips and dipterans. Ingke Rachor (Wien) and Martin Schmidt (Bern) provided essential help concerning the field experiment. The Deutsche Forschungsgemeinschaft financially supported this study.

References

Agusti N, Shayler P, Harwood JD, Vaughan IP, Sunderland KD, Symondson WOC (2003) Collembola as alternative prey sustaining spiders in arable ecosystems: prey detection within predators using molecular markers. Molecular Ecology 12:3467-3475

Albers D (2002) Nahrungsnetz und Stoffdynamik auf extensiv bewirtschafteten Ackerflächen - die Untersuchung stabiler Isotope (13C, 15N) im Zersetzer-Subsystem. PhD thesis, Göttingen

Blüthgen N, Gebauer G, Fiedler K (2003) Disentangling a rainforest food web using stable isotopes: dietary diversity in a species-rich ant community. Oecologia 137:426-435

Brüning A (1991) The effect of a single colony of the red wood ant, *Formica polyctena*, on the spider fauna (Araneae) of a beech forest floor. Oecologia 86: 478-483

Cabana G, Rasmussen JB (1994) Modelling food chain structure and contaminant bioaccumulation using stable nitrogen isotopes. Nature 372:255–257

Cronin JT, Haynes KJ, Dillemuth F (2004) Spider effects on planthopper mortality, dispersal, and spatial population dynamics. Ecology 85: 2134-2134

DeNiro MJ, Epstein S (1978) Influence of diet on the distribution of carbon isotopes in animals. Geochim Cosmochim Acta 42:495–506

DeNiro MJ, Epstein S (1981) Influence of diet on the distribution of nitrogen isotopes in animals. Geochim Cosmochim Acta 45:341–351

Denno RF, Mitter MS, Langellotto GA, Gratton C, Finke DL (2004) Interactions between a hunting spider and a web-builder: consequences of intraguild predation and cannibalism for prey suppression. Ecol Entomol 29:566-577

Ende CN von (1993) Repeated-measures analysis: growth and other time-dependent measures. In: Scheiner SM, Gurevich J (eds) The Design and Analysis of Ecological Experiments. Oxford University Press, Oxford, pp134–157

Finke DL, Denno RF (2003) Intra-guild predation relaxes natural enemy impacts on herbivore populations. Ecol Entomol 28: 67-73

Finke DL, Denno, RF (2004) Predator diversity dampens trophic cascades. Nature 429: 407-410

Gibb H (2003) Dominant meat ants affect only their specialist predator in an epigaeic arthropod community. Oecologia 136:609-615

Halaj J, Ross DW, Moldenke AR (1997) Negative effects of ant foraging on spiders in Douglas-fir canopies. Oecologia 109: 313-322

Hölldobler B, Wilson E0 (1995) The ants. Springer, Berlin Heidelberg New York

Kajak A, Breymeyer A, Pętal J (1971) Productivity investigation of two types of meadows in the vistula valley. IX Predatory arthropods. Ekol Pol 19: 223-233

Kajak A, Breymeyer A, Pętal J, Olechowicz E (1972) The influence of ants on the meadow invertebrates. Ekol Pol 20: 163-171

Kempson D, Lloyd M, Ghelardi R (1963) A new extractor for woodland litter. Pedobiologia 3: 1-21

Kling GW, Fry B, O'Brien WJ (1992) Stable isotopes and planktonic trophic structure in Arctic lakes. Ecology 73:561-566

Laakso J, Setälä H (2000) Impact of wood ants (*Formica aquilonia* Yarr.) on the invertebrate food web of the boreal forest floor. Ann Zool Fennici 37: 93-100

Lajtha K, Michener RH (eds) (1994) Stable isotopes in ecology and environmental science. Blackwell, Oxford

Lang A (2003) Intraguild interference and biocontrol effects of generalist predators in a winter wheat field. Oecologia 134:144-153

Lawrence KL, Wise DH (2000) Spiders predation on forest-floor Collembola and evidence for indirect effects on decomposition. Pedobiologia 44:33-39

Lawrence KL, Wise DH (2004) Unexpected indirect effect of spiders on the rate of litter disappearance in a deciduous forest. Pedobiologia 48:149-151

Lenoir L, Bengtson J, Persson T (2003) Effects of Formica ants on the soil fauna - results from a short-term exclusion and a long-term natural experiment. Oecologia 143:423-430

Macko SA, Lee WY, Parkere PL (1982) Nitrogen and carbon fractionation by two species of marine amphipods: laboratory and field studies. J Exp Mar Biol Ecol 63:145–149

Magnusson WE, Carmozina de Araújo M, Cintra R, Lima AP, Martinelli LA, Sanaiotti TM, Vasconcelos HL, Victoria RL (1999) Contributions of C3 and C4 plants to higher trophic levels in an Amazonian savanna. Oecologia 119:91–96

McNabb DM, Halaj J, Wise DH (2001) Inferring trophic positions of generalist predators and their linkage to the detrital food web in agroecosystems: a stable isotope analysis. Pedobiologia 45:289-287

Minagawa M, Wada E (1984) Stepwise enrichment of 15-N along food chains: Further evidence and the relation between 15-N and animal age. Geochimi Cosmochimi Acta 48:1135-1140

Nentwig W (1986) Non-webbuilding spiders: prey specialists or generalists. Oecologia 69:571-576

Nyffler M (1999) Prey selection of spiders in the field. Journal of Arachnology 27:317-324

Oelbermann K, Scheu S (2002) Stable isotope enrichment ($\delta^{15}N$ and $\delta^{13}C$) in a generalist predator (Pardosa lugubris, Araneae: Lycosidae): effects of prey quality. Oecologia 130:337–344

Oraze MJ, Grigarick AA (1989) Biological control of aster leafhopper (Homoptera: Cicadellidae) and midges (Diptera: Chironomidae) by Pardosa ramulosa (Araneae: Lycosidae) in California rice fields. Journal of Economic Entomology 82:745-749

Otto D (1965) Der Einfluss der Roten Waldameise (*Formica polyctena* Först.)auf die Zusammensetzung der Insektenfauna (ausschließlich gradierende Arten). Collana Verde 16:250-263

Owens NJP (1987) Natural variations in 15N in the natural environment. Adv Mar Biol 24:389–451

Pętal J, Breymeyer A (1969) Reduction off wandering spiders by ants in a Stellario-Deschampsietum Meadow. Bull Acad Pol Sci Cl. II 17:239-244

Petelle M, Haines B, Haines E (1979) Food preferences analyzed using 13C/12C Ratios. Oecologia 38:159-166

Peterson BJ, Fry B (1987) Stable isotopes in ecosystem studies. Annu Rev Ecol Syst 18:293–320

Pitkin BR (1988) Lesser dung flies, Diptera: Sphaeroceridae. Royal entomological Society of London, London

Polis GA, Myers CA, Holt RD (1989) The ecology and evolution of intraguild predation: potential competitors that eat each other. Annu Rev Ecol Syst 20:297-330

Ponsard S, Arditi R (2000) What can stable isotopes (d15N and d13C) tell about the food web of soil macroinvertebrates. Ecology 81:852-864

Post DM (2002) Using stable isotopes to estimate trophic position: models, methods, and assumptions. Ecology 83:703-718

Reineking A, Langel R, Schikowski J (1993) 15-N, 13-C-on-line measurements with an elemental analyzer (carlo erba, NA 1500), a modified trapping box and a gas isotope mass spectrometer (Finnigan, MAT 251). Isotopenpraxis 29:169-174

Riechert SE, Bishop L (1990) Prey control by an assemblage of generalist predators: Spiders in an Garden test systems. Ecology 71:1441-1450

Riechert SE, Lawrence K (1997) Test for predation effects of single versus multiple species of generalist predators: spiders and their insect prey. Entomol Exp Appl 84:147-155

Rosenheim JA, Wilhoit LR, Armet CA (1993) Influence of intraguild predation among generalist insect predators on the suppression of herbivore population. Oecologia 96:439-449

Schauermann J (1982) Verbesserte Extraktion der terrestrischen Bodenfauna im Vielfachgerät modifiziert nach Kempson und MacFadyen. Kurzmitteilungen aus dem SFB 135 (Ökosysteme auf Kalkgestein) 1:47-50

Schmidt MH, Lauer A, Purtauf T, Thies C, Schaefer M, Tscharntke T (2003) Relative importance of predators and parasitoids for cereal aphid control. Proc R Soc Lond B 270:1905–1909

Schmitz OJ (1998) Direct and indirect effects of predation and predation risk in old-field interaction webs. Am Nat 151:327-342

Seifert B (1996) Ameisen: beobachten, bestimmen. Naturbuch-Verlag, Augsburg

Sigsgaard (2002) Early season natural biological control of insect pests in rice by spiders - and some factors in the management of the cropping system that may affect this control:57-64 In: European Arachnology 2000 (S. Toft & N. Scharff eds). Aarhus University Press, Århus

Smith KGV (1989) An introduction to the immature stages of British flies, Diptera larvae, with notes on eggs, puparia and pupae. Royal entomological Society of London, London

Snyder WE, Wise DH (2001) Contrasting trophic cascades generated by a community of generalist predators. Ecology 82:1571-1583

Stein TM (1996) Klimabeobachtungen in Witzenhausen für das Jahr 1995. Arbeiten und Berichte Nr. 43 des FG Kulturtechnik und Ressourcenschutz, University of Kassel

Van der Aart P, de Wit T (1971) A field study on interspecific competition between ants (Formicidae) and hunting spiders (Lycosidae, Gnaphosidae, Ctenidae, Pisauridae, Clubionidae) Neth J Zool 21:117-126

Van der Zanden MJ, Rasmussen JB (1999) Primary consumer $\delta 13C$ and $\delta 15N$ and the trophic position of aquatic consumers. Ecology 80:1395–1404

Vanderklift MA, Ponsard S (2003) Sources of variation in consumer-diet $\delta 15N$ enrichment: a meta analysis. Oecologia 136:169-182

Wada E, Mizutani H, Minagawa M (1991) The use of stable isotopes for food web analysis. Crit Rev Food Sci Nutr 30:361-371

Wise DH (1993) Spiders in ecological webs. Cambridge University Press, Cambridge

Wise DH (2004) Wandering spiders limit densities of a major microbi-detritivore in the forest-floor food web. Pedobiologia 48:181-188

Chapter 3

Chapter 3

Predator diversity and top down effects: Intraguild interactions with hunting spiders reduce top-down control of web-builders

Dirk Sanders and Christian Platner

Abstract

The loss in species diversity demands deeper insights into predator-prey interactions in food webs and into the function of diversity. The increasing species diversity of generalist predators may enhance the strength of top down effects, due to different strategies in catching prey. However, intraguild interactions among predators can reduce their ability of prey suppression.

In field experiments we tested the single and combined predatory effects of web-building spiders and hunting spiders on the arthropod community of a grassland. These experiments were conducted for two diversity levels: one containing one species of web-building and hunting spiders; the other containing three. Natural spider densities were established inside fenced plots and manipulated by the removal of spiders during the three months of the experiment.

In comparison to the single species systems, the effects of spiders on lower trophic levels were stronger in the more diverse predator treatment. Auchenorrhyncha densities were 25 % lower in plots with web-building spiders. This effect of web-building spiders was reduced by the presence of hunting spiders.

The analysis of stable isotopes ^{13}C and ^{15}N revealed a higher trophic position in the food web for hunting spiders than for web-builders and also emphasised the occurrence of intraguild predation. In contrast, web-builders seemed to feed predominantly on herbivores.

In high predator diversity treatments biomass of plants was 20 % higher than in low diversity treatments. This indicates the positive effect of a more diverse and abundant predator guild. The density of large-sized springtails (Collembola) was reduced by 30 % in plots with hunting spiders.

The more diverse predator guild also contained more individuals, so stronger effects for the more diverse spider assemblage were not surprising. However, if intraguild predators such as hunting spiders were included, the per capita effects and top-down effects on Auchenorrhyncha population declined with increasing predator abundance and diversity.

Keywords Field experiment, generalist predators, leafhoppers, planthoppers, stable isotopes

Introduction

Declining biodiversity and its implications for continued provision of ecosystem services have led to an intense research effort to study the relationships between biodiversity and ecosystem functioning (Loreau et al. 2001, Wilby & Thomas 2002, Duffy 2003). Predators can strongly control herbivore populations, which can be an important ecosystem service regarding agricultural systems. Unfortunately, predators are more susceptible to local and regional extinctions than species at other trophic levels (Duffy 2002, 2003). A change in the diversity of predators is known to affect the strength of trophic cascades (Finke & Denno 2004, Snyder et al. 2006). Spiders are potential generalist predators with regard to prey suppression in natural (Riechert & Bishop 1990, Wise 1993) and agricultural systems (Symondson, Sunderland & Greenstone 2002, Schmidt et al. 2003). A combination of different strategies of catching prey (species complementarity) by increasing the species diversity of predators can enhance their ability for prey suppression (Riechert & Bishop 1990, Riechert & Lawrence 1997, Snyder et al. 2006). However, intraguild interactions among predators can reduce this effect (Snyder and Wise 2001, Lang 2003, Arim & Marquet 2004, Finke & Denno 2003, 2004, Denno et al. 2004).

Field experiments are an important method for studying trophic interactions and predatory effects under natural conditions (Wise 1993, Hodge 1999). In our study, in order to detect if different hunting strategies complement one another and, thus, result in stronger top down effects, we tested the single and combined effect of the two functional groups of web-builders and hunting spiders on insect populations in a grassland system. The field experiment was conducted for two diversity levels containing natural densities of either one spider species, or three spider species of each functional group. Two dominant spiders at the study site were chosen for the single predator treatment: the web-builder *Argiope bruennichi* (Scopoli) (Araneidae) and the hunting spider *Pisaura mirabilis* (Clerck) (Pisauridae). A wolf spider and a thomisid species were added to the hunting spiders' treatment, and an agelenid and a theridiid species were added to the web-builders' treatment, thus resulting in three species systems.

For potential prey groups, we chose planthoppers and leafhoppers (Auchenorrhyncha: Fulgoromorpha and Cicadomorpha) as the dominating herbivores of our study site. These insects generally account for a high proportion of the biomass and species diversity in most grasslands, and are highly responsive to changes in their environment (e.g. Waloff 1980; Biedermann et al. 2005). Collembola (Springtails) are an important prey group for ground-living spiders (Wise 2004, Sanders & Platner 2007). In addition, we used stable isotopes ^{13}C and ^{15}N, which is a promising method for studying trophic links in food web analysis (De Niro & Epstein 1981,Wada, Mizutani & Minagawa 1991, Kling, Fry & O'Brien 1992, Ponsard & Arditi 2000, Wise, Moldenhauer & Halaj 2006). It can also provide important information that explains the response of arthropod groups in a field experiment (Sanders & Platner 2007). Predatory effects on prey groups and, indirectly, on plant biomass in this study may reveal if (1) different strategies in catching prey result in stronger top down control or (2) intraguild interactions reduce top down forces.

Materials and methods

Study site

The research was conducted in the experimental area of the Faculty of Agriculture at the University of Goettingen (Lower Saxony, Germany). This area comprised of a fallow with stands of quack grass (*Agropyron repens* L.) and creeping bent grass (*Agrostis stolonifera* L.) where the plots were established.

Pisaura mirabilis, *Pardosa amentata* (Clerck) and *Xysticus* spec. were abundant hunting spiders in this system, with 20-30 individuals/m^2 combined. Biomass-dominant web-building spiders with densities of up to 6 individuals/m^2 were: the orb web spider *Argiope bruennichi*; the funnel-web spider *Agelena gracilis* (Koch); and the tangle web spider *Enoplognatha ovata* (Clerck). The herbivorous guild in the grassland consisted mainly of planthoppers and leafhoppers, with *Mocydia crocea* (Herrich-Schäffer), *Arthaldeus pascuellus* (Fallén), *Streptanus aemulans* (Kirschbaum) and *Delphacodes venosus* (Germar) being the dominant species.

Experiment

The basic experimental unit was a 1 m^2 area, enclosed by a 50 cm high fence of gauze. The fence was applied on a 20 cm plastic ring, which was buried 10 cm deep into the ground. The experiment ran from 3rd June until 28th August 2005, and was set up in a two-factorial design, the two factors being the "hunting spider" and the "web-building spider". Natural densities of these two functional groups of spiders were established inside selected plots and removed from the remaining plots. We established this experiment in two levels of spider diversity within the functional groups: one species and three species systems. There were seven treatments: one treatment without spiders and two treatments testing the effects of a single species of each functional group on its own and one treatment for both species combined. This was done in the same way for the three species assemblages in three more treatments (see Table 1 for specific species composition). All treatments were replicated five times in blocks, giving a total of 35 plots. In each of the blocks an additional Auchenorrhyncha-removal treatment was established to simulate a strong predatory effect, and to assess the response of plant biomass. As a control, in order to judge the effects of the enclosures, one reference sample was taken from each of the five blocks outside the plots in similar vegetation.

At the start of the experiment the vegetation was cut to a height of 15 cm. This was done for two reasons: on the one hand, manipulation of the fauna is easier in shorter vegetation; on the other hand, the quack grass grows fast and we wanted to asses the predator effects on plant biomass from the start of the experiment. The plots were defaunated with a suction sampler, and after vacuuming, invertebrates were again released into the plots, excluding spiders in spider-removal-plots and Auchenorrhyncha in the Auchenorrhyncha-removal-treatment. For the one-species systems we chose *P. mirabilis* and *A. bruennichi*. In the three-species treatment, natural densities of the large web-building spiders, *A. bruennichi*, *A. gracilis* and *E. ovata* were established, while we used *P. amentata*, *P. mirabilis* and *Xysticus* spec. for the hunting spiders' treatment.

The abundance of spiders was higher in the three-species systems, because we established natural densities of each spider species. Therefore, the increase in the species diversity of the predator guild cannot be separated from the simultaneous increase in predator abundance. We decided to run the experiment in this way to simulate a more natural situation, as opposed to diversity experiments in which the density of predators remained at the same level while the diversity was increased (Snyder et al. 2006). In natural systems, an extinct predator may not be replaced by other individuals of the remaining species. In our experiment, in the single species treatment, the density of spiders was lower than in control samples, while spider densities were natural for the assemblages. Equal predator densities for all diversity treatments would have resulted in unnaturally high densities of single spider species or lower densities for the assemblages (Fig. 1).

The predator-removal treatment was achieved by removing spiders manually twice a week during the three months of the experiment. Each plot was searched for spiders by one person for five minutes and, on average, 2 to 4 (mainly *Pardosa*) spiders per plot were removed from the removal-plots.

Table 1 Experimental design of the field experiment with eight treatments (replicated five times). The two factors "hunting spider" and "web-building spider" were conducted in two diversity levels (single species and three species systems). An Auchonorrhyncha-removal treatment was established to simulate a strong predatory effect and to asses the response of plant biomass. For control of the effect of enclosures, a reference sample was taken outside the plots in similar vegetation.

Treatment (5 replicates)	Diversity Level	Hunting spiders	Web-builders	Initial densities [Ind./m^2]
Removal	-	-	-	0
Hunting spider (H)	1	Pisaura	-	4
Web-builder (W)	1	-	Argiope	3
H + W	1	Pisaura	Argiope	7
Hunting spiders	3	Pisaura, Pardosa, Xysticus	-	16
Web-builders	3	-	Argiope, Agelena, Enoplognatha	8
H + W	3	Pisaura, Pardosa, Xysticus	Argiope, Agelena, Enoplognatha	24
Auchenorrhyncha removal		Not manipulated		
Unfenced control		Not manipulated		

Sampling

The fauna was sampled with a suction sampler (Stihl SH 85, Germany; 10 s/sample using a 0.036 m² sampling cylinder) at the end of the experiment. For each plot an area of 0.18 m^2 was sampled. Spiders, planthoppers and leafhoppers were identified to species level, while other arthropods were assigned to higher-ranking taxa. In order to estimate plant biomass, plants were cut from an area of 0.03 m2 in the centre of each plot. The plants were dried for 72 h at a temperature of 60°C, and the dry weight was measured. As a control, and to judge the effect of manipulation, spider densities were estimated by searching for spiders inside the plots once a week in July and August.

Stable isotopes

Ratios of ^{13}C and ^{15}N were estimated by a coupled system consisting of an elemental analyzer (Carlo Erba NA 2500) and a gas isotope mass spectrometer (Finnigan Deltaplus). The system was computer-controlled, allowing measurement of ^{13}C and ^{15}N (Reineking, Langel & Schikowski 1993). Isotopic contents were expressed in δ units as the relative difference between sample and conventional standards with $δ^{15}N$ or $δ^{13}C$ [‰] = (RSample – RStandard)/RStandard x 1000, where R is the ratio of $^{15}N/^{14}N$ or $^{13}C/^{12}C$ content, respectively. The conventional standard for ^{15}N is atmospheric nitrogen and for ^{13}C PD-belemnite (PDB) carbonate (Ponsard & Arditi 2000). Acetanilide (C8H9NO, Merck, Darmstadt) served for internal calibration with a mean standard deviation of samples <0.1‰. Samples were dried for 72 h (60°C) and weighed into tin capsules to contain 500-1800 µg of dry biomass. We analysed spiders, their potential prey and plants.

Data analyses

The effect of the treatments and the response of the arthropod community were analysed by a two factor analysis of variance (ANOVA procedure, SAS version 8). The general linear model (GLM procedure, SAS) was used to compare the effects of predator treatment and Auchenorrhyncha-removal on Auchenorrhyncha and plant biomass, when the data were unbalanced. All abundance and biomass data were log-transformed to meet assumptions of normality and homogeneity of variances.

Results

Manipulation

In removal plots no individuals of the three large web-building spider species were found. *Argiope* was present with on average 3 individuals/m^2 in the single-species-treatment for web-building-spiders (Fig. 1, Table 2), while the web-building spider assemblage in the three-species-treatments had densities with 6 individuals/m^2 (Fig. 1, Table 2). The abundance of web-builders declined from July to August (Fig. 1). Small, juvenile linyphiid spiders were removed at the start of the experiment but were able to re-colonize the plots and reached densities of up to 120 indivuals/m^2 at the end, with no differences concerning the treatments.

The density of *Pisaura* in the one species treatment was 10 times lower than in the non-removal plots and the density of the hunting spider assemblage was 16 times lower in removal plots than in the non-removal plots (Fig. 1, Table 2). *Pisaura* was present with 5 individuals/m^2 in the one species treatment, and in the assemblage of hunting spiders, with 16 individuals/m^2 in the three species treatment (Fig. 1). The density of planthoppers and leafhoppers was reduced by 60% in Auchenorrhyncha-removal plots (Fig 2. GLM $F_{1,19}$ = 6.51; p = 0.0195).

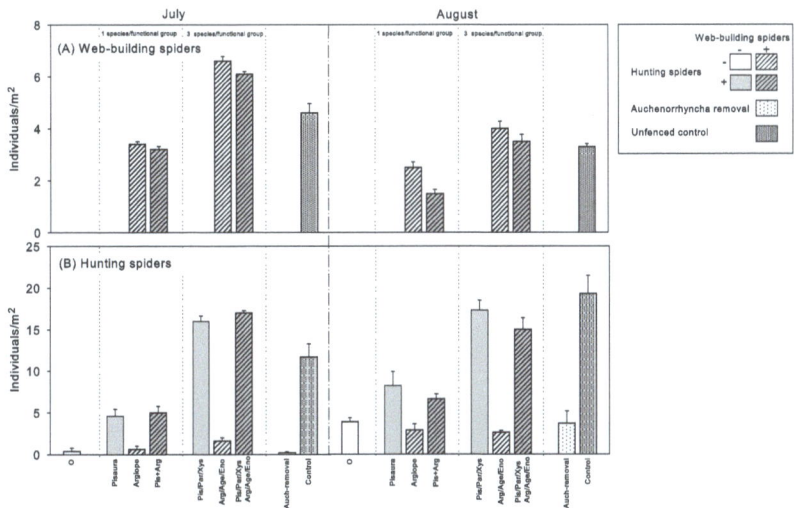

Fig. 1 Mean abundance (+1SE) of (A) web-building spiders and (B) hunting spiders in July and August in the eight different treatments with natural and reduced hunting spider and web-building spider density with two levels of predator diversity (one and three species systems) and in Auchenorrhyncha-removal-plots and control samples from outside the plots. Open bars (0): plots with reduced spiders density; shaded bars: plots with natural hunting spider density; hatched bars: plots with natural web-building spider density; dotted bars: Auchenorrhyncha-removal; dotted shaded bars: controls from outside the plots.

Table 2 Response of hunting spiders and web-building spiders to the treatments. For control of the effect of manipulation, spider densities were estimated by searching for spiders inside the plots once a week in July and August. Data were log-transformed (log10X+1). F values are given for a repeated measures ANOVA for mean densities from July and August; for the within effects F values for Pillai's Trace are given. df = degrees of freedom (Nom, Den), bold digits indicates statistical significance ($p<0.05$).

	df	Web-building spiders		Hunting spiders	
		F	P	F	P
Web	1, 12	3807.21	**<.0001**	1.44	0.2534
Hunt	1, 12	9.95	**0.0083**	392.87	**<.0001**
W × H	1, 12	9.95	**0.0083**	0.28	0.6072
Block	4, 12	0.82	0.5379	0.79	0.5556
Time	1, 12	27.44	**0.0002**	178.74	**<.0001**
T × W	1, 12	27.44	**0.0002**	8.70	**0.0121**
T × H	1, 12	0.02	0.8947	41.80	**<.0001**
T×W×H	1, 12	0.02	0.8947	0.32	0.5826
T × Bl	4, 12	0.60	0.6669	2.47	0.1011

Top down control

In single-species treatments, no effects of spiders on lower trophic levels were found (Fig. 2, Table 3). However, in the three species treatment, planthoppers and leafhoppers responded strongly to the presence of web-building spiders. Densities were 25% lower when compared to the spider-removal plots (Fig. 2, Table 3). The effects of hunting spiders on planthoppers and leafhoppers, as well as the effects of both of the spider groups combined, were not significant (Fig. 2, Table 3). The interaction of the factors "hunting spiders x web-builders", however, was significant (Table 3).

The biomass of plants tended to respond positively to Auchenorrhyncha-removal, which served to simulate the predatory effects on herbivores, but without statistical significance (Fig. 2; ANOVA $F_{1,8}$ = 2.46; p= 0.1557). However, plant biomass was positively affected by the presence of predators (GLM $F_{1,19}$ = 4.79; p = 0.0413) and by a more diverse predator guild. In the three species treatments, plant biomass was 19% higher than in the single-species systems (Fig. 2, ANOVA $F_{1,24}$ = 4.46; p = 0.0454). No effects of predator diversity were found on Auchenorrhyncha density (ANOVA $F_{1,24}$ = 0.20; p = 0.6577). However, the diversity of Auchenorrhyncha was negatively affected by the three species treatment. Average species richness in the three species treatment was 4.9, and in the single species treatment, 5.7 (ANOVA $F_{1,24}$ = 4.27; p = 0.0499). The density of large-sized springtails (Collembola) was, on average, 30 % lower in the presence of hunting spiders in the three species treatment (Fig. 3; Table 3). Collembola abundance was lower inside the enclosures in comparison to control samples ($F_{1,38}$ = 5.05, P =0.0305, for GLM).

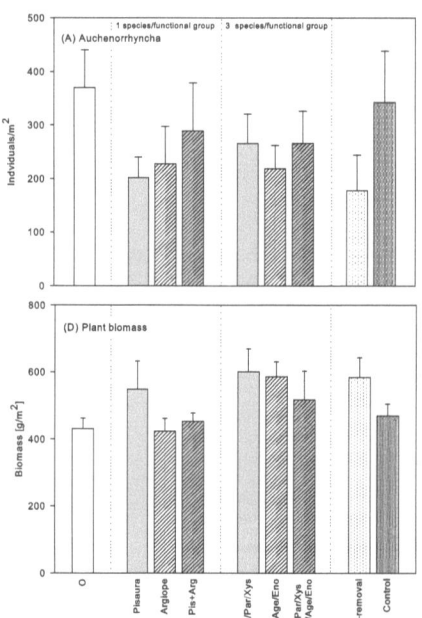

Fig. 2 Mean abundance (+1SE) of (A) Auchenorrhyncha and (B) plant biomass in August in the eight different treatments with natural and reduced hunting spider and web-building spider density with two levels of predator diversity (one and three species systems) and in Auchenorrhyncha-removal-plots and control samples from outside the plots.

The per capita impact of predators on the Auchenorrhyncha population [log(Auchenorrhyncha density in absence of predators/ Auchenorrhyncha density in presence of predators)/predator density in July] was estimated and, because of the additive design (Finke & Denno 2005), was used to correct the differences in the abundances of predators across the predator compositions. The per capita impact of spiders on Auchenorrhyncha was highest for the hunting spiders in the single-species treatment, and lowest for the assemblage of hunting spiders and web-building spiders in the three species treatment (Table 4). In comparison to the treatment with both web-builders and hunting spiders present, the impact was generally higher in treatments with only one functional group of spiders, although this was marginally significant ($F_{1,28}$ = 3.67,

Table 3 Response of Auchenorrhyncha, plant biomass and Collembola for single-species systems and three-species systems using a two-way ANOVA. Data were log-transformed (log10X+1). df = degrees of freedom, bold digits indicates statistical significance (p<0.05).

	df	Web-building spiders 1, 12		Hunting spiders 1, 12		W x H 1, 12		Block 4, 12	
		F	p	F	p	F	p	F	p
Single species treatment									
Auchenorrhyncha		0.47	0.5065	0.62	0.4472	3.14	0.1019	1.17	0.3736
Collembola		1.18	0.2996	0.56	0.4682	0.24	0.6315	2.76	0.0775
Plant biomass		0.84	0.3766	2.27	0.1575	0.42	0.5312	2.33	0.1151
Three species treatment									
Auchenorrhyncha		7.42	**0.0185**	1.37	0.2651	6.00	**0.0306**	17.73	**<.0001**
Collembola		0.22	0.6454	4.46	**0.0563**	1.04	0.3274	4.25	**0.0227**
Plant biomass		0.28	0.6058	0.44	0.5175	4.33	0.0595	0.71	0.6027

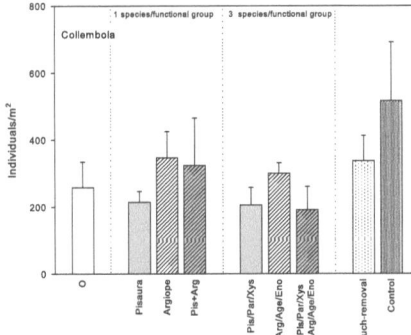

Fig. 3 Mean abundance (+1SE) of Collembola (springtails) in August in the eight different treatments with natural and reduced hunting spiders and web-building spider density with two levels of predator diversity (one and three species systems) and in Auchenorrhyncha-removal-plots and control samples from outside the plots.

P =0.0657, for GLM). *Per capita* impacts of spiders in the single species treatment were significantly higher than in the three-species treatment ($F_{1,28}$ = 6.92, P =0.0137, for ANOVA).

Food web analysis

Plants had a $\delta^{13}C$ value of –28.5‰ and $\delta^{15}N$ value of 3.2 ‰ (Fig. 4). Auchenorrhyncha were enriched in $\delta^{15}N$ compared to plants with a $\delta^{15}N$ value of 4.2 ‰. Collembola were also more enriched in $\delta^{15}N$ than plants. The spiders most enriched in ^{15}N were the hunting spiders, i.e. *Pardosa amentata* and *Pisaura mirabilis*, with $\delta^{15}N$ values of 8‰. $^{15}N/^{14}N$ ratios in web-building spiders (*Argiope*, *Agelena*, *Enoplognatha*) were significantly lower than in hunting spiders (Fig. 4, $F_{1,9}$ = 36.56, P = <0.0001, for GLM). The heteropteran bugs, Nabidae, Miridae and Lygaeidae, had $\delta^{15}N$ values similar to web-building spiders.

Table 4 *Per capita* impact of predators on Auchenorrhyncha population in the different treatments [log(Auchenorrhyncha density in absence of predators/ Auchenorrhyncha density in presence of predators)/predator density in July] *Tukey* test for statistical significance (p<0.05).

	Treatment	Per capita impact Mean	SE
Single species	Hunting spiders	0.113	± 0.027a
	Web-builders	0.075	± 0.033ab
	H+W	0.032	± 0.021ab
Three species	Hunting spiders	0.024	± 0.014ab
	Web-builders	0.036	± 0.012ab
	H+W	0.014	± 0.009 b

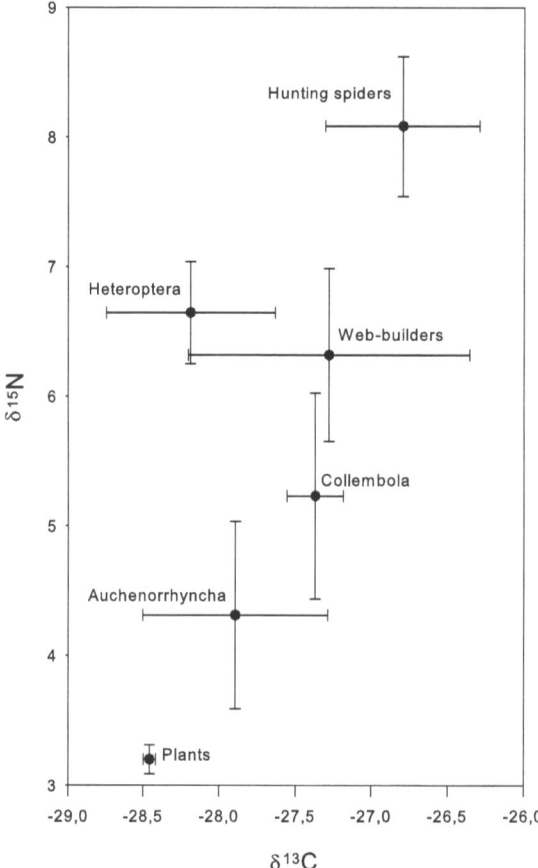

Fig. 4 $\delta^{15}N$ and $\delta^{13}C$ values (± SE) of hunting spiders, web-building spiders, Heteroptera (Miridae, Nabidae, Lygaeidae) Auchenorrhyncha and Collembola. Samples were taken from plots of the field experiment and replicated five times for each taxon.

Discussion

Manipulation

We successfully manipulated the densities of large web-builders and hunting spiders. Small juvenile linyphiid spiders could not be excluded for the entirety of the experiment, however, these juvenile spiders are not able to build large webs and are known to feed predominantly on collembolans (Sanders & Platner 2007). All spiders that were large enough to be visually detected were removed two times a week. In the simulation of the predation pressure on Auchenorrhyncha by removal with a suction sampler, juvenile members of Auchenorrhyncha were probably able to remain in the litter layer during defaunation of the plots, and the extent of the effects on the plant biomass is not based on a reduction of all Auchenorrhyncha individuals.

Top down control

In the single-species treatments, we found no statistical evidence for a strong top down control of spiders. In contrast, the assemblage of the three web-building spider species strongly affected plant- and leafhopper abundance, where the density of Auchenorrhyncha was 25 % lower compared to predator-removal plots. However, this effect was reduced by the presence of hunting spiders. We assume that intraguild interactions took place and relaxed prey suppression of web-building spiders in these plots. In addition, *per capita* effects of spiders were reduced in treatments with both spider groups present. The stable isotope data demonstrated that hunting spiders had a position one trophic level higher than web-builders (Fig. 4), indicating that their food resource include members of their own guild. Intraguild predation and cannibalism are widespread in arthropod-dominated communities (Polis, Meyers & Holt 1989, Rosenheim, Wilhoit & Armet 1993, Wise 1993) and are known for hunting spiders of the genus *Pisaura*, *Pardosa* and *Xysticus* (Nentwig 1986). These kinds of interference can strongly affect the strength of top down control (Finke & Denno 2003) and weaken tropic cascades (Finke & Denno 2004). Web-builders in our system were strict insectivores (Nyffeler 1999) – this is also supported by the stable isotope data (Fig. 4) – and seemed to be more important for top-down control on herbivores. Heteropteran bugs had a similar trophic position to web-building spiders (Fig. 4), indicating that they also feed on lower trophic levels. It is probable that they are important predators of Auchenorrhyncha, which was, however, not tested in our experiment. Therefore, only the web-builders which are stricter insectivores were able to affect the Auchenorrhyncha population strongly in the higher diversity treatment. Reducing the predator guild to one species resulted in unnaturally low predator densities, which were not able to exert strong effects on the Auchenorrhyncha density, although the *per capita* impact was higher for single species than for the assemblage, especially for *Pisaura*. Hunting spiders as intraguild predators negatively affected the top down control of web-builders due to intraguild predation.

There was evidence for a trophic cascade generated by spiders in our system, although this was not strong for a single functional group (web-builders or hunting spiders). The Auchenorrhyncha removal resulted in an increase in plant biomass to a degree comparable to predator effects, but this effect was not statistically significant. A comparison between the removal plots and the predator plots demonstrated a positive influence of predators on plant biomass. In the three-species treatments, biomass of plants was 20 % higher as compared to single-species systems, indicating a trophic cascade initiated by the spider assemblage. Trophic cascades by generalist predators could be demonstrated for terrestrial ecosystems (Moran & Hurd 1998, Schmitz, Hämback & Beckermann 2000), but they are not as strong as in aquatic or marine systems (Halaj & Wise 2001, Shurin et al. 2002). However, diversity treatment had no effect on Auchenorrhyncha density, although the average species diversity of Auchenorrhyncha declined in the more diverse predator treatment.

A diverse predator guild appears to be able to affect lower trophic levels more than less diverse species systems, although *per capita* impacts of predators were stronger for the single species treatment. In our experiment, a higher diverse predator guild also contained more individuals, so it is not surprising that the effects were stronger for the more diverse spider assemblage. In contrast, if intraguild predators such as hunting spiders were included, the *per capita* effects and top-down effects on the Auchenorrhyncha population declined with increasing predator abundance and diversity. This may be an important implication for biocontrol of pest species in agro-ecosystems. In biological control it is desired to increase plant biomass and crop yield by suppressing density of herbivores. Our results suggest that top-down control is stronger in systems with predator assemblages containing in majority stricter insectivores.

Acknowledgments

We are grateful to David Wise and Matthias Schaefer for their valuable discussions about the manuscript. Thanks are due to Mari Annina Gough and Sharon Cooper for their linguistic help. Thomas Grützner and Anna Gilles provided essential help concerning the field experiment. The Agraroekologie Goettingen was so kind as to provide the experimental area. This study was financially supported by the Deutsche Forschungsgemeinschaft.

References

Arim, M. & Marquet, P.A. (2004) Intraguild predation: a widespread interaction related to species biology. Ecology Letters, 7, 557–564.

Biedermann, R., Achtziger, R., Nickel H. & Stewart A.J.A. (2005) Conservation of grassland leafhoppers: an introductory review. Journal of Insect Conservation, 9, 229–243.

DeNiro, M.J. & Epstein, S. (1981) Influence of diet on the distribution of nitrogen isotopes in animals. Geochimica et Cosmochimica Acta, 45, 341–351.

Denno, R.F., Mitter, M.S., Langellotto, G.A., Gratton, C. & Finke, D.L. (2004) Interactions between a hunting spider and a web-builder: consequences of intraguild predation and cannibalism for prey suppression. Ecological Entomology, 29, 566–577.

Duffy, J.E. (2002) Biodiversity and ecosystem function: the consumes connection. Oikos, 99, 201–219.

Duffy, J.E. (2003) Biodiversity loss, trophic skew and ecosystem functioning. Ecology Letters, 6, 680–687.

Finke, D.L. & Denno R.F. (2003) Intra-guild predation relaxes natural enemy impacts on herbivore populations. Ecological Entomology, 28, 67–73.

Finke, D.L. & Denno, R.F. (2004) Predator diversity dampens trophic cascades. Nature, 429, 407–410.

Finke, D.L. & Denno, R.F. (2005) Predator diversity and the functioning of ecosystems: the role of intraguild predation in dampening trophic cascades. Ecology Letters, 8, 1299–1306.

Halaj, J. & Wise, D.H. (2001) Terrestrial trophic cascades: How much do they trickle? American Naturalist, 157, 262–281.

Hodge, M.A. (1999) The implications of intraguild predation for the role of spiders in biological control. The Journal of Arachnology, 27, 351–362.

Kling, G.W., Fry, B. & O'Brien, W.J. (1992) Stable isotopes and planktonic trophic structure in Arctic lakes. Ecology, 73, 561–566.

Lang, A. (2003) Intraguild interference and biocontrol effects of generalist predators in a winter wheat field. Oecologia, 134,144–153.

Loreau, M., Naeem, S., Inchausti, P., Bengtsson, J., Grime, J.P., Hector, A., Hooper, D.U., Huston, M.A., Raffaelli, D., Schmid, B., Tilman, D. & Wardle, D.A. (2001) Biodiversity and ecosystem functioning: Current knowledge and future challenges. Science, 294, 804–808.

Moran, M.D. & Hurd, L.E. (1998) A trophic cascade in a diverse arthropod community caused by a generalist arthropod predator. Oecologia, 113, 126–132.

Nentwig, W. (1986) Non-webbuilding spiders: prey specialists or generalists. Oecologia, 69, 571–576.

Nyffeler, M. (1999) Prey selection of spiders in the field. Journal of Arachnology, 27, 317–324.

Polis, G.A., Myers, C.A. & Holt, R.D. (1989) The ecology and evolution of intraguild predation: potential competitors that eat each other. Annual Review of Ecology and Systematics, 20, 297–330.

Ponsard, S. & Arditi, R. (2000) What can stable isotopes ($d^{15}N$ and $d^{13}C$) tell about the food web of soil macroinvertebrates. Ecology, 81, 852–864.

Reineking, A., Langel, R. & Schikowski, J. (1993) 15-N, 13-C-on-line measurements with an elemental analyzer (carlo erba, NA 1500), a modified trapping box and a gas isotope mass spectrometer (Finnigan, MAT 251). Isotopenpraxis, 29, 169–174.

Riechert, S.E. & Bishop, L. (1990) Prey control by an assemblage of generalist predators: Spiders in a garden test systems. Ecology, 71, 1441–1450.

Riechert, S.E. & Lawrence, K. (1997) Test for predation effects of single versus multiple species of generalist predators: spiders and their insect prey. Entomologia Experimentalis et Applicata, 84, 147–155.

Rosenheim, J.A., Wilhoit, L.R. & Armet, C.A. (1993) Influence of intraguild predation among generalist insect predators on the suppression of herbivore population. Oecologia, 96, 439–449.

Sanders, D., Nickel, H., Grützner, T. & Platner, C. (in press) Habitat structure mediates top-down effects of spiders and ants on herbivores. Basic and Applied Ecology.

Sanders, D. & Platner, C. (2007) Intraguild interactions between spiders and ants and top-down control in a dry grassland. Oecologia, 150, 611–624.

Schmidt, M.H., Lauer A., Purtauf, T., Thies, C., Schaefer, M. & Tscharntke, T. (2003) Relative importance of predators and parasitoids for cereal aphid control. Proceedings of the Royal Society of London, 270, 1905–1909.

Schmitz, O.J., Hämback, P.A., Beckerman, A.P. (2000) Trophic cascades in terrestrial systems: a review of the effects of carnivore removals on plants. American Naturalist, 155, 141–153.

Shurin, J.B., Borer, E.T., Seabloom, E.W., Anderson, K., Blanchette, C.A., Broitman, B., Cooper, S.D. & Halpern, B.S. (2002) A cross-ecosystem comparison of the strength of trophic cascades. Ecology Letters, 5, 785–791.

Snyder, W.E., Snyder, G.B., Finke, L.F. & Straub, C.S. (2006) Predator biodiversity strengthens herbivore suppression. Ecology Letters, 9, 789–796.

Snyder, W.E., Wise, D.H. (2001) Contrasting trophic cascades generated by a community of generalist predators. Ecology, 82,1571–1583.

Symondson, W.O.C., Sunderland, K.D. & Greenstone, H.M. (2002) Can generalist predators be effective biocontrol agents? Annual Review of Entomology, 47, 561–594.

Wada, E., Mizutani, H. & Minagawa, M. (1991) The use of stable isotopes for food web analysis. Critical Reviews in Food Science and Nutrition, 30, 361–371.

Waloff, N. (1980) Studies on grassland leafhoppers and their natural enemies. Advances in Ecological Research, 11, 81–215.

Wilby, A. & Thomas, M.B. (2002) Natural enemy diversity and pest control: patterns of pest emergence with agricultural intensification. Ecology Letters, 5, 353–360.

Wise, D.H. (1993) Spiders in ecological webs. Cambridge University Press, Cambridge.

Wise, D.H. (2004) Wandering spiders limit densities of a major microbi-detritivore in the forest-floor food web. Pedobiologia, 48, 181–188.

Wise D.H., Moldenhauer, D.M. & Halaj, J. (2006) Using stable isotopes to reveal shifts in prey consumption by generalist predators. Ecological Applications, 16, 865–876.

Chapter 4

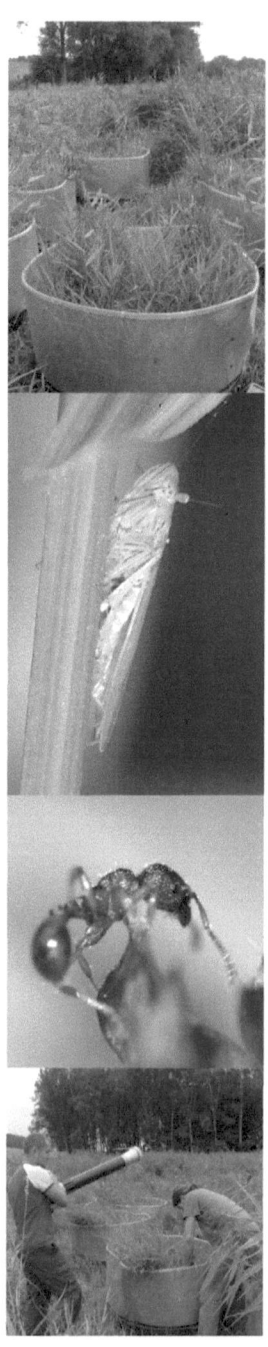

Chapter 4

Test for effects of functional diversity: ants, hunting spiders and web-builders in a wet grassland food web

Abstract

Generalist predators such as ants and spiders are highly abundant in most terrestrial ecosystems, thus stressing their importance for ecosystem processes. We studied the community effects of web-building spiders, hunting spiders and ants in a three-factorial designed field experiment, which was carried out for two years in a wet grassland habitat using fenced plots with an inside area of one square metre. Density of predators was manipulated by continuous removal of spiders and ants nests. The response of the arthropod community which belong mostly to herbivores and detritivores was assessed by suction sampling. In addition, plant biomass was estimated.

Population size of the most abundant planthopper species *Erzaleus metrius* was reduced by 50 % in the presence of hunting spiders and web-builders. However, two other important herbivores, the heteropteran bug *Ischnodemus sabuleti* and the planthopper *Stenocranus major*, were not affected by predator removal. Similarly, there was no evidence for a trophic cascade to plant biomass generated by the three predator groups. Collembola (springtails) were negatively affected by hunting spiders and web-builders, as well as by ants in the second year, indicating a strong trophic link between generalist predators and the detritivores. The density of harvestmen declined in the presence of web-building spiders, indicating intraguild interference. Microbial biomass was positively correlated with the density of herbivores. The analysis of stable isotopes ^{13}C and ^{15}N revealed a higher trophic position in the food web for hunting spiders than for web-builders and ants due to intraguild predation, while web-builders and ants seem to feed predominantly on lower trophic levels.

In conclusion, we found no evidence for an additive effect of the three generalist predator groups regarding their top-down control. Additionally, they seem to depend mainly on Collembola as a food resource, while predatory effects on the herbivores were not strong enough to be passed on to plants.

Keywords Field experiment, generalist predators, leafhoppers, planthoppers, stable isotopes

Introduction

Predators can provide a key ecosystem service by regulating herbivore populations in agricultural and other systems. Unfortunately, predators are more susceptible to local and regional extinctions than species at other trophic levels (Duffey 2002, 2003). Thus the dominant impacts of biodiversity change on ecosystem functioning appear to be trophically mediated (Duffey 2003). Ants and spiders as generalist predators occur in large numbers in most terrestrial ecosystems (Wise 1993; Hölldobler and Wilson 1995), which underlines their importance for ecosystem processes as lying-in-wait predators, that can switch to pest species in times of their mass occurrence and are effectively able to control herbivore populations (Riechert & Bishop 1990, Wise 1993, Symondson, Sunderland & Greenstone 2002, Schmidt et al. 2003). However, generalist predators also frequently feed on other predators, and such intraguild interactions can reduce the possibility of prey control (Rosenheim et al. 1993, Snyder and Wise 2001, Lang 2003, Arim & Marquet 2004, Finke & Denno 2003, 2004, Denno et al. 2004). A combination of different strategies of catching prey by increasing the diversity of predators can enhance the ability of predator guilds for prey suppression (Riechert & Lawrence 1997, Snyder et al. 2006). Spiders can be divided into two main functional groups according to their strategies for catching prey. Web-builders belonging to various families employ silk to assist in the capture of prey. Among the hunting spiders some lie motionless in ambush and are typical sit-and-wait predators, while others actively go in search for their prey (e.g. Lycosidae, Pisaura, Salticidae). Ants are also hunters able to prey on large arthropods by recruiting nest mates, which largely extends the range of possible prey. Studies have reported high predation rates by ants of the genus *Myrmica* on arthropods in a meadow (Petal and Breymeyer 1969; Kajak et al. 1972). Predators can affect both the herbivores (Schmitz et al. 2000, Halaj and Wise 2001) and the detritivorous system, and change decomposition processes (Lawrence and Wise 2004, Lensing and Wise 2006). In our study, in order to detect if different hunting strategies complement one another and, thus, result in stronger top down effects, we tested the single and combined effect of the three functional groups of ants, web-builders and hunting spiders on arthropod populations in a field experiment. Field experiments are an important method for studying trophic interactions and predatory effects under natural conditions (Wise 1993, Hodge 1999). We chose a natural wet grassland system for the experiment to test ecological hypotheses that could yet be demonstrated in mesocosm experiments or more simple systems (Finke and Denno 2005, Snyder et al. 2006). In addition to the experiment, we used the analysis of stable isotopes ^{13}C and ^{15}N, which is a promising method for studying trophic links in food web analysis (De Niro & Epstein 1981, Wada, Mizutani & Minagawa 1991, Kling, Fry and O'Brien 1992, Ponsard & Arditi 2000, Wise, Moldenhauer & Halaj 2006). It can also provide important information that explains the response of arthropod groups in a predator-removal experiment (Sanders and Platner 2007).

In the current study, we manipulated densities of hunting spiders, web-building spiders and ants in a field experiment and tested their effects as predators on an arthropod community. Predatory effects on prey groups and, indirectly, on plant biomass in this study may reveal (1) which predator group has the strongest predatory impact on lower trophic levels, and (2) if the combination of different strategies in catching prey results in a stronger top down control.

Materials and methods

Study site

The field experiment was conducted in a wet meadow near Göttingen (Lower Saxony, Germany) with stands of the reed canarygrass (*Phalaris arundinacea* L.) and common reed (*Phragmites australis* L.).

Ants of the two species *Myrmica rubra* (L.) and *M. ruginodis* (Nylander) were present with on average one colony/m^2 at the study site. The dominant spider group were the web-builders with densities up to 1000 individuals/m^2. Larger sheet webs were built by the linyphiids *Neriene clathrata* (Sundevall), *Floronia bucculenta* (Clerck) and *Tenuiphantes tenuis* (Blackwall), and three *Bathyphantes* species with densities for these linyphiids of up to 16 individuals/m^2. The tangle web spider *Neottiura bimaculata* (L.) reached high densities with 300 juveniles/m^2. Hunting spiders (e.g. *Pisaura mirabilis*, *Pardosa amentata* (Clerck), thomisid genera *Xysticus* and *Ozyptila* and Clubionidae) were less abundant than the web-builders with on average 90 individuals/m^2. The herbivorous guild in the grassland consisted mainly of planthoppers and leafhoppers, with *Erzaleus metrius* (Flor), and *Stenocranus major* (Kirschbaum) being the dominant species and the heteropteran species *Ischnodemus sabuleti* (Fallén).

Experiment

The basic experimental unit was a 1 m^2 area, enclosed by a 50 cm high fence of gauze. At the top margin of the fence a sliced colourless tube was applied, which was brushed with Vaseline as a slippery barrier for exclusion of spiders and ants. The fence was fixed on a 20 cm high plastic ring, which was buried 10 cm deep into the ground. The experiment ran from 3rd Mai 2006 until 30th August 2007, and was set up in a three-factorial design, the factors being the "hunting spider", the "web-building spider" and the "ant". Natural densities of these three groups of generalist predators were established inside selected plots, and the individuals were removed from the remaining plots. All treatments were replicated four times in blocks, giving a total of 32 plots. In each of the blocks an additional Hemiptera-removal treatment was established to simulate a strong predatory effect, and to assess the response of plant biomass. As a control, in order to judge the effects of the enclosures, one reference sample was taken at each of the four blocks outside the plots from an area with similar vegetation compared to the plots.

At the start of the experiment the plots were defaunated with a suction sampler, and after vacuuming, invertebrates were again released into the plots, excluding spiders in spider-removal-plots and Auchenorrhyncha and Heteroptera in the Hemiptera-removal-treatment. The low predator-density treatment was achieved indirectly by placing slippery barriers on the outside of the rings and by removing spiders manually and excluding ant colonies. Spider populations and ant colonies that became re-established in the removal plots were removed manually. Each plot was searched for spiders and ant nests by one person for ten minutes and, on average, 15-20 mainly juvenile spiders of the therridiids and linyphiids per plot were removed from the removal-plots. Ant colonies were excavated and replaced by soil cores containing the same vegetation compared to the plot. The control was done once a week in spring, summer and autumn, while in winter the plots remained unmanipulated.

Sampling

The fauna was sampled with a suction sampler (Stihl SH 85, Germany; 10 s/sample using a 0.036 m² sampling cylinder) in early summer and at the end of August. The sampling cylinder was attached to the ground and exhausted with the suction sampler. This was repeated four times in each plot, which resulted in an area of 0.14 m². Spiders, ants and planthoppers and leafhoppers were identified to species level, while other arthropods were assigned to higher-ranking taxa. Suction sampling took place in June and August/September in 2006 and 2007, resulting in four sampling dates. In order to estimate plant biomass, plants were cut from an area of 0.03 m² in the centre of each plot in the first year, and the entire vegetation was cut in the second year at the end of the experiment. The plants were dried, and the dry weight was measured. As a control, and to judge the effect of manipulation, spider webs were counted using a spray bottle to make the webs visible.

Stable isotopes

Ratios of ^{13}C and ^{15}N were estimated by a coupled system consisting of an elemental analyzer (Carlo Erba NA 2500) and a gas isotope mass spectrometer (Finnigan Delta^{plus}). The system was computer-controlled, allowing measurement of ^{13}C and ^{15}N (Reineking, Langel & Schikowski 1993). Isotopic contents were expressed in δ units as the relative difference between sample and conventional standards with $δ^{15}N$ or $δ^{13}C$ [‰] = ($R_{Sample} - R_{Standard}$)/$R_{Standard}$ x 1000, where R is the ratio of $^{15}N/^{14}N$ or $^{13}C/^{12}C$ content, respectively. The conventional standard for ^{15}N is atmospheric nitrogen and for ^{13}C PD-belemnite (PDB) carbonate (Ponsard & Arditi 2000). Acetanilide (C_8H_9NO, Merck, Darmstadt) served as internal calibration with a mean standard deviation of samples <0.1‰. Samples were dried for 72 h (60°C) and weighed into tin capsules to contain 500-1800 µg of dry biomass. We analysed spiders, ants, their possible prey and plants.

Data analyses

The effects of factors being the "hunting spider", the "web-building spider" and the "ant", and the response of the arthropod community, were analysed by a repeated measures three factor analysis of variance (rmANOVA) (Ende 1993) with a repeated time of four sampling occasions. For plant biomass and for webs of spiders we had data for only two sampling occasions in September 2006 and August 2007. All abundance and biomass data were log-transformed to meet assumptions of normality and homogeneity of variances.

Results

Table 1 Response of ants, hunting spiders and web-building spiders to the treatments. Data were log-transformed (log10X+1). F values are given for a repeated measures ANOVA for suction samples from June and September in 2006 and 2007; for the within effects F values for Pillai's Trace are given. df = degrees of freedom (Nom, Den), bold digits indicates statistical significance (p<0.10).

	df	Myrmica F	Myrmica P	Web-builders F	Web-builders P	Hunting spiders F	Hunting spiders P
Web-builder (W)	1, 21	0.14	0.7180	6.32	**0.0201**	0.79	0.3852
Hunting spider (H)	1, 21	25.88	**<0.0001**	0.01	0.9251	4.11	**0.0555**
Ant (A)	1, 21	26.83	**<0.0001**	4.66	**0.0426**	0.22	0.6402
W × H	1, 21	16.07	**0.0006**	0.60	0.4485	0.34	0.5683
W × A	1, 21	4.80	**0.0398**	0.19	0.6666	0.16	0.6930
H × A	1, 21	0.76	0.3940	0.75	0.3968	0.10	0.7563
W × H × A	1, 21	3.17	0.0894	0.19	0.6658	0.18	0.6737
Block	3, 12	1.21	0.3315	0.86	0.4784	0.30	0.8258
Time (T)	3, 19	1.59	0.2244	176.23	**<0.0001**	49.51	**<0.0001**
T × W	3, 19	0.98	0.4212	1.04	0.3981	1.38	0.2787
T × H	3, 19	1.81	0.1796	0.81	0.5033	0.99	0.4182
T × A	3, 19	0.22	0.8844	2.35	0.1042	0.15	0.9273
T × W × H	3, 19	0.77	0.5224	0.46	0.7129	5.13	**0.0091**
T × W × A	3, 19	0.38	0.7692	0.71	0.5569	0.56	0.6510
T × H × A	3, 19	0.42	0.7442	0.50	0.6842	1.66	0.2095
T × W × H × A	3, 19	0.33	0.8026	1.02	0.4042	0.76	0.5277
T × Block	9, 63	0.39	0.9353	1.47	0.1795	1.60	0.1362

Manipulation

Ants of the genus *Myrmica* were present with one colony and a density measured by suction sampling with 20-30 individuals/m^2 in the plots. Some colonies had to be removed due to the immigration of complete ant populations. The count of spider webs proved a successful manipulation of the larger sheet web-building-spiders (Fig. 1, Rep meas. ANOVA $F_{1;21}$= 174.11; p <0.0001). The abundance of web-builders, mainly theridiids of *Neottiura bimaculata*, was also significantly affected by the manipulation (Table 1), although juveniles remained in removal plots with on average densities of 500 individuals/m^2 compared to 800 individuals/m^2 in web-building spider plots (Fig. 1). The hunting spiders had a density of 80 individuals/m^2 in the hunting spider plots and the control samples from outside the plots. Their density could be reduced in removal plots to densities of 40-60 individuals/m^2 (Fig. 1, Table 1). We found a positive effect of ants on the population size of web-building spiders (Table 1, Fig. 1), which increased in density by 23% in plots with *Myrmica* colonies. The density of ants was likewise positively affected by the presence of hunting spiders (Table 1).

In Hemiptera-removal, population size of most abundant herbivores such as *Ischnodemus sabuleti* (Heteroptera) (ANOVA $F_{1;3}$ = 21.50; p = 0.0189) and the planthopper species *Erzaleus metrius* and *Stenocranus major* (ANOVA $F_{1;3}$ = 9.97.; p = 0.0510) could only be reduced in the second year of the study (from 240±40 to 80±30 individuals/m^2).

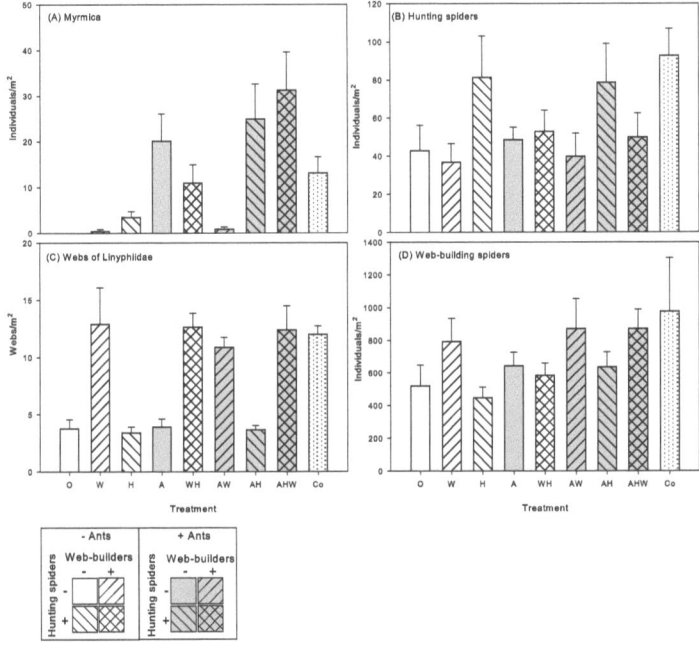

Fig. 1 Mean abundance (+1SE) of ants (A) hunting spiders (B), webs of Linyphiidae (C) and web-building spiders (d) in the eight different treatment combinations with natural and reduced ant, hunting spiders and web-builder density and in control samples outside the plots (Co). Open bars: plots with reduced spiders and ant density; shaded bars: plots with natural ant density and hatched bars refer to plots with natural spiders density; dotted bars: controls from outside the plots. For statistical analyses see text and table 1.

Top down control

Population size of the planthopper species *Erzaleus metrius* was negatively affected by the presence of hunting spiders and web-building spiders (Fig. 2, Table 2). The density declined by 50% in plots with spiders, an effect that was reduced by the presence of ants, however, predator diversity had no effect on the density of *Erzaleus* (GLM Test for the effect of predator diversity $F_{2;25}= 0.70$, p = 0.5077). The heteropteran bug *Ischnodemus sabuleti* and the planthopper species *Stenocranus major* were not influenced by predator removal (Fig 2, Table 2). In contrast to spiders, ants had no effect on herbivore population at all. Biomass of plants responded negatively to ant treatment (Rep meas. ANOVA $F_{1;21}= 2.98$; p = 0.0990) and was increased from 900±90 g/m² in the control to 1280±80 g/m² in the Hemiptera-removal at the end of the experiment (ANOVA $F1;3= 53.52$, p = 0.0053).

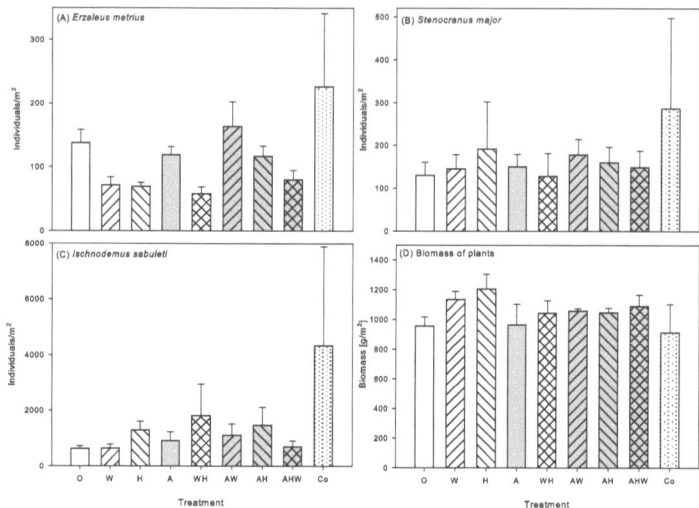

Fig. 2 Mean abundance (+1SE) of the planthopper species *Erzaleus metrius* (A), and *Stenocranus major* (B) and the heteropteran bug *Ischnodemus sabuleti* (C) and biomass of plants (D) in the eight different treatment combinations with natural and reduced ant, hunting spiders and web-builder density and in control samples outside the plots (Co). Open bars: plots with reduced spiders and ant density (0); shaded bars: plots with natural ant density (A) and hatched bars refer to plots with natural spiders density (W and H); dotted bars: controls from outside the plots (for a legend see Fig. 1). For statistical analyses see text and table 2.

Table 2 Response of members of the arthropod community from suction samples to the manipulation of ants, hunting spider and web-building spider density. All comparisons were made using a three-way repeated measures ANOVA for data from June and September in 2006 and 2007; for the within effects F values for Pillai's Trace are given. Data were log-transformed (log10X+1). df = degrees of freedom (Nom, Den), bold digits indicates statistical significance ($p<0.05$).

	df	Stenocranus major		Erzaleus metrius		Ischnodemus sabuleti		Plant biomass	
		F	P	F	P	F	P	F	P
Web-builder (W)	1, 21	0.06	0.8034	4.24	0.0520	1.66	0.2118	1.58	0.2228
Hunting spider (H)	1, 21	0.04	0.8509	4.99	**0.0365**	0.04	0.8510	0.00	0.9892
Ant (A)	1, 21	0.73	0.4030	3.38	0.0801	0.77	0.3894	2.98	0.0990
W × H	1, 21	1.26	0.2744	0.09	0.7682	3.29	0.0838	1.04	0.3194
W × A	1, 21	0.34	0.5646	0.50	0.4869	0.03	0.8726	0.05	0.8207
H × A	1, 21	0.11	0.7439	0.13	0.7234	0.85	0.3665	1.59	0.2214
W × H × A	1, 21	0.00	0.9755	3.89	0.0619	0.07	0.7973	2.21	0.1519
Block	3, 31	5.57	**0.0057**	0.79	0.5104	0.65	0.5896	2.24	0.1134
Time (T)	3, 19	81.8	**<0.0001**	171.6	**<0.0001**	103.19	**<0.0001**	4.65	**0.0429**
T × W	3, 19	1.76	0.1892	0.17	0.9161	0.49	0.6963	0.16	0.6903
T × H	3, 19	1.46	0.2566	0.74	0.5424	1.19	0.3396	2.10	0.1621
T × A	3, 19	4.41	**0.0163**	0.74	0.5397	1.52	0.2405	1.09	0.3081
T × W × H	3, 19	0.69	0.5680	0.79	0.5131	0.48	0.6970	1.11	0.3042
T × W × A	3, 19	5.99	**0.0047**	1.35	0.2895	0.70	0.5612	0.91	0.3515
T × H × A	3, 19	0.95	0.4363	0.49	0.6938	14.46	**<0.0001**	2.02	0.1697
T × W × H × A	3, 19	0.39	0.7639	0.53	0.6666	0.92	0.4483	0.20	0.6621
T × Block	9, 63	1.38	0.2166	1.35	0.2302	2.15	**0.0381**	3.53	**0.0324**

Collembola, which were larger than 1 mm, responded to the presence of web-building spiders with a reduction in population size by 34% (Table 3, Fig. 3), while small sized Collembola were negatively affected by hunting spider treatment (Table 3, Fig. 3). This effect was reduced in treatments with ants and with all three predator groups (Table 3, see significant interactions). Ant treatment had a significantly positive effect on small sized Collembola density by 31% in September of the first year (ANOVA $F_{1;21}= 9.11$, $p = 0.0065$). However, Collembola density declined by 36 % in plots with ants in the second year (ANOVA $F_{1;21}= 4.99$, $p = 0.0364$). Diversity of predators had no effect on population size of Collembola (GLM Test for the effect of predator diversity $F_{2;25}= 0.63$, $p = 0.5390$). (GLM Test for the effect of predator diversity $F_{3;28}= 3.69$, $p = 0.0235$). The density of harvestmen (Opiliones) was reduced by 20 % in the presence of web-building spiders (Table 3, Fig. 3).

Table 3 Response of members of the arthropod community from suction samples to the manipulation of ants, hunting spider and web-building spider density. All comparisons were made using a three-way repeated measures ANOVA for data from June and September in 2006 and 2007; for the within effects F values for Pillai's Trace are given. Data were log-transformed (log10X+1). df = degrees of freedom (Nom, Den), bold digits indicates statistical significance (p<0.05).

	df	Opiliones		Collembola large sized		Collembola small sized	
		F	P	F	P	F	P
Web-builder (W)	1, 21	8.10	**0.0073**	6.32	**0.0202**	1.83	0.1906
Hunting spider (H)	1, 21	1.10	0.3065	0.01	0.9081	7.00	**0.0151**
Ant (A)	1, 21	0.00	0.9956	0.02	0.8967	0.61	0.4439
W × H	1, 21	0.30	0.5895	0.40	0.5317	1.02	0.3229
W × A	1, 21	0.00	0.9731	2.54	0.1263	2.88	0.1045
H × A	1, 21	1.38	0.2525	2.78	0.1101	7.19	**0.0139**
W × H × A	1, 21	6.20	**0.0213**	0.00	0.9492	6.79	**0.0165**
Block	3, 31	1.32	0.2929	9.37	**0.0004**	0.38	0.7649
Time (T)	3, 19	74.20	**<0.0001**	8.91	**0.0007**	93.46	**<0.0001**
T × W	3, 19	4.18	**0.0197**	0.85	0.4857	0.20	0.8968
T × H	3, 19	5.22	**0.0085**	6.38	**0.0036**	0.42	0.7389
T × A	3, 19	0.52	0.6750	1.71	0.1988	3.84	**0.0265**
T × W × H	3, 19	3.52	**0.0352**	0.34	0.7970	0.35	0.7914
T × W × A	3, 19	0.63	0.6055	0.80	0.5094	0.93	0.4466
T × H × A	3, 19	5.29	**0.0080**	2.71	0.0740	0.86	0.4786
T × W × H × A	3, 19	1.33	0.2934	0.41	0.7492	0.87	0.4716
T × Block	9, 63	0.53	0.8471	2.89	**0.0064**	1.33	0.2385

Food web analysis

The dominant plant species at the experimental area *Phalaris arundinacea* had a $\delta^{13}C$ value of – 25.6 ± 0.2 ‰ and a $\delta^{15}N$ value of 2.1± 0.1 ‰ (Fig. 4). Most analyzed arthropods had similar $\delta^{13}C$ values compared to the grass. Diplopoda and Isopoda were slightly more enriched in ^{13}C (-24.4±0.3 and −24.5±0.4 ‰, respectively). Collembola and Blattoptera had similar $\delta^{15}N$ values with 2.9±0.5 and 3.0±0.1 ‰. Herbivores were more enriched in ^{15}N, such as the planthoppers species *Erzaleus metrius* (4.1±0.5 ‰) and *Stenocranus major* (4.7±0.7 ‰), and $\delta^{13}C$ values of −26.7±0.3

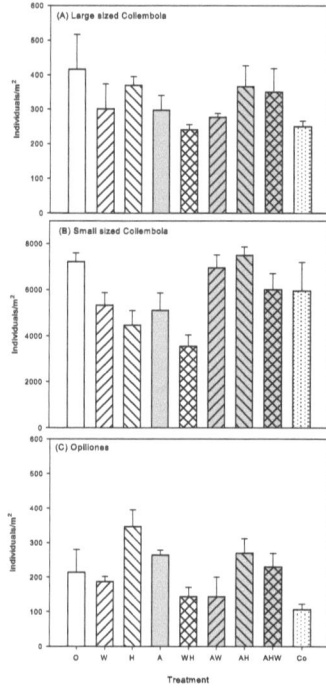

Fig. 3 Mean abundance (+1SE) of large Collembola (A), small Collembola (B) and Opiliones (C) in the eight different treatment combinations with natural and reduced ant, hunting spiders and web-builder density and in control samples outside the plots (Co). Open bars: plots with reduced spiders and ant density (0); shaded bars: plots with natural ant density (A) and hatched bars refer to plots with natural spiders density (W and H); dotted bars: controls from outside the plots (for a legend see Fig. 1). For statistical analyses see text and table 2.

‰ and −25.9±0.3 ‰, respectively. *Ischnodemus sabuleti* is monophagous on *Phalaris arundinacea* with a very similar δ^{13}C (−25.4±0.4 ‰) compared to the plant and a δ^{15}N value of 6.0±0.3 ‰. The linyphiid spiders *Bathyphantes* (δ^{15}N 5.8±0.6 ‰), *Floronia bucculenta* (δ^{15}N 6.2±0.2 ‰), *Neriene clathrata* (δ^{15}N 7.2±0.4 ‰) and *Tenuiphantes tenuis* (δ^{15}N 7.2±0.4 ‰) had an intermediate position in the food web, which was similar for ants of the genus *Myrmica* with a δ^{15}N of 6.9±0.8 ‰ (Fig. 5). Theridiids of the species *Neottiura bimaculata* were less enriched with δ^{13}C than the other spiders with −26.3±0.1 ‰ and had a δ^{15}N value of 7.3±0.4 ‰. Hunting spiders were, in general, more enriched with ^{15}N in comparison to web-builders. *Zora* had the lowest δ^{15}N value of hunting spiders with a δ^{15}N of 7.3±0.4 ‰, which is similar to that of *Tenuiphantes* and *Neriene*. The spiders most enriched with ^{15}N were *Pardosa amentata* (8.8±0.7 ‰), *Clubiona* (8.6±0.02 ‰) and *Pisaura mirabilis* (8.1±0.3 ‰). The mimitid spider *Ero*, which is known to feed on web-building spiders, had a similar ^{15}N content (8.2±0.7 ‰) compared to hunting spiders. Opiliones had similar stable isotope values to spiders (δ^{15}N 7.7±0.2 and δ^{13}C −25.1±0.2 ‰).

Discussion

We found that all three generalist predator groups had a negative impact on the population size of Collembola (Fig. 6). Web-builders seemed to feed mainly on large sized Collembola, while hunting spiders depressed the density of smaller Collembola. However, this effect of hunting spiders was reduced by the synchronous presence of ants in the plots. For ants the effects were different in the two years of the study. We found a significant positive effect on Collembola density, which increased in plots with ants by 31% in September of the first year. In contrast, Collembola density declined by 36 % in the second year. Ants of the species *Myrmica rubra* are known to feed on Collembola and are also able to switch to springtails as mass prey (Reznikova and Panteleeva 2001). Dissimilarly, in an other study we found a positive effect of mounds of the ant *Lasius niger* on the abundance of springtails in a grassland in spring (Schuch et al., submitted). Hence, both positive and negative influences are in the rage of possible interactions between ants and springtails. We assume that the extreme variation may at least partly reflect

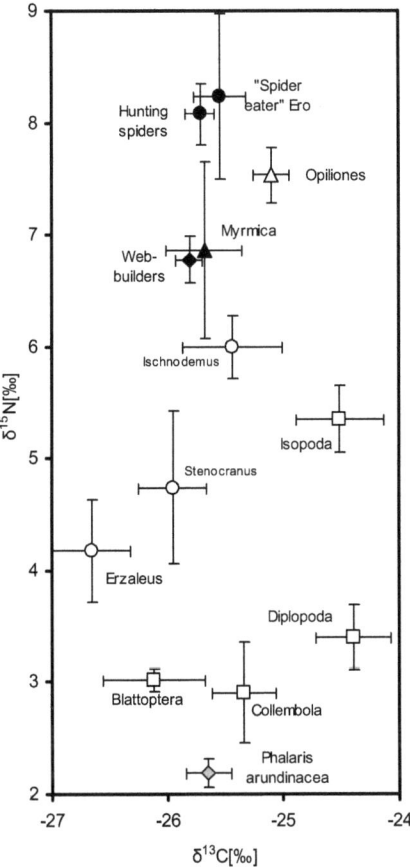

Fig. 4 $\delta^{15}N$ and $\delta^{13}C$ values (± SE) of spiders, ants and their possible prey organisms and of plants (shaded diamond). ● = hunting spiders, ♦ = web-building spiders, ▲ = ants, ○ = herbivores, □ = detritivorous and fungivorous groups.

differences in abiotic factors regarding the two years. The unexpected positive effects of ants on Collembola occurred during a period of unusually warm and dry weather in the first year, which contrasts to a wet and cold summer in the second year. Probably springtails could benefit from the activity of ants, which is known to modify chemical and physical soil properties (Dauber and Wolters 1999, Frouz et al. 2003, Dostál et al. 2005, Lane and BassiriRad 2005). Generalist predators in agro-ecosystems often depend on members of the detritivorous subsystem as a food resource (Scheu 2001, Sigsgaard 2002; Agusti et al. 2003), which was also demonstrated for hunting spiders in a natural grassland (Sanders and Platner 2007).

In contrast to Collembola the overall herbivore density was not strongly affected by the three generalist predator groups. Only one of the three most abundant herbivore species, e.g. Erzaleus metrius, responded to the presence of hunting spiders and web-builders with a decline in density by 50%. Further, there was no evidence for a trophic cascade generated by generalist predators. However, biomass of plants was higher in the Hemiptera-removal treatment, which served as simulation of a strong predatory effect on herbivores. This demonstrated that, in our system, it was not the linkage between the herbivores and plants that was weak as stated in a review by Shurin et al. (2002), but the linkage between predators and herbivores. This may be explained by the presence of the highly abundant heteropteran species Ischnodemus sabuleti that seem to be avoided by the predators. Scheu (2001) postulated a prey switching of generalist predators in a dry summer period from detritivores to herbivores in arable systems due to the movement of the detritivores in lower soil layers. Decreased moisture of a forest floor in an experiment reduced overall Collembola density, but led to a higher Collembola activity (Shultz et al. 2006). Probably the water content of the soil in the wet grassland was high enough in summer to maintain the soil organism in the surface and there was no need for the generalist predators to switch to herbivores as prey, especially if they are not as "tasty" with regard to Ischnodemus having scent glands. We conclude

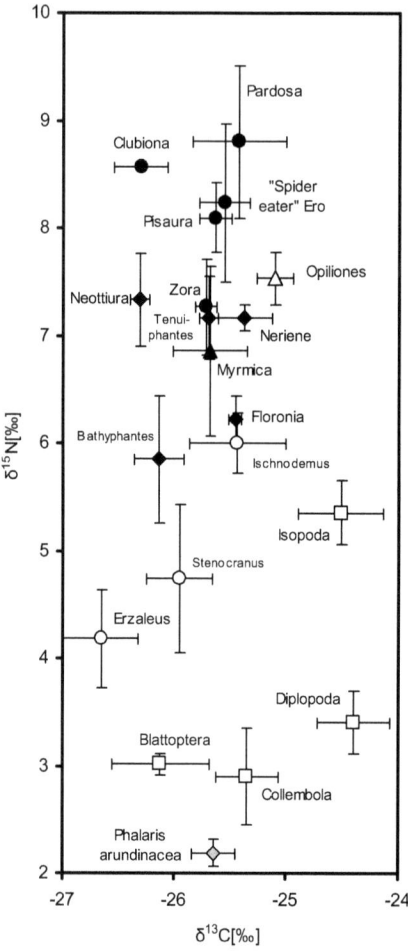

Fig. 5 Detailed presentation of $\delta^{15}N$ and $\delta^{13}C$ values (± SE) of spiders, ants and their possible prey organisms and of plants (shaded diamond).. ● = hunting spiders, ♦ = web-building spiders, ▲ = ants, ○ = herbivores, □ = detritivorous and fungivorous groups.

that Collembola, which declined in population size in the presence of spiders and ants, are the main food resource of the generalist predators in the wet grassland.

Intraguild predation is a common phenomenon in generalist predator guilds (Arim and Marquet 2004). Our stable isotope data demonstrate that hunting spiders, which have similar $\delta^{15}N$ values compared to the "spider eater" *Ero*, are intraguild predators. Web-builders contain less ^{15}N indicating that they feed mainly on lower trophic levels. These results are similar to another study focussing on the intraguild interactions between web-builders and hunting spiders (Sanders *et al.* submitted). *Myrmica* had a similar intermediate trophic position in the food web as such as web-builders, which demonstrates that their food also is manly derived from lower trophic levels. Interference among the predator groups was not statistically proven, which indicates that such interactions only occur among hunting spiders, or do not affect populations on community level. However, if a predator group reaches higher densities in plots without the other groups this may indicate a situation of interference or competition. This pattern was observed for hunting spiders, which occur in plots without web-builders with an abundance of 80 individuals/m² and declined in the presence of web-builders to 50 individuals/m². In contrast, there was a significant positive effect of ants on the population size of web-building spiders, which were 23% more abundant in plots with ants. We observed web-building spiders feeding on *Myrmica*, but the stable isotope data state clearly that ants are not a main food resource. This positive effect may be explained by an indirect effect via the positive influence of springtails. We observed a negative response of harvestmen (Opiliones) to web-building spider treatment. Harvestman, *i.e.* the most abundant species *Nemastoma lugubre*, also feed on small insects, and a competitive exclusion based on the reduction of large sized Collembola by web-building spiders seems to be more likely than a predatory interaction.

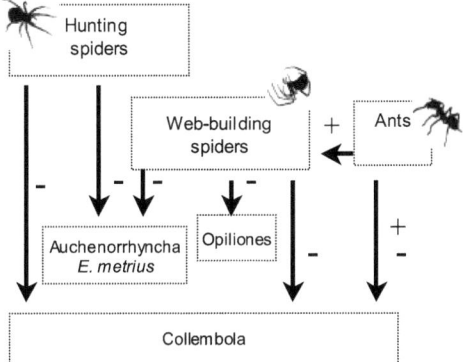

There was no evidence for a stronger top down control of predator guilds containing three functional groups. However, in the treatment with both hunting spiders and web-builders, prey suppression of herbivores and detritivores was highest. Ants seem to have a special role in the food web, interacting positively with other predators and possible prey.

Fig. 6 Model of trophic interactions in the studied food web.

Acknowledgments

We are grateful to Mari Annina Gough for her valuable discussion about the manuscript. Sharon Cooper and Simone König provided essential help during the field experiment. This study was financially supported by the Deutsche Forschungsgemeinschaft.

References

Arim, M. & Marquet, P.A. (2004) Intraguild predation: a widespread interaction related to species biology. *Ecology Letters*, 7, 557–564.

Boyer A. G., Swearingen R. E., Blaha M A., Fortson C T., Gremillion S K., Osborn K A. Moran M D. (2003) Seasonal variation in top-down and bottom-up processes in a grassland arthropod community. *Oecologia* 136:309–316.

Dauber J, Wolters V, 1999. Microbial activity and functional diversity in the mounds of three different ant species. *Soil Biology & Biochemistry* 32, 93-99.

DeNiro, M.J. & Epstein, S. (1981) Influence of diet on the distribution of nitrogen isotopes in animals. *Geochimica et Cosmochimica Acta*, 45, 341–351.

Dostál P, Březnová M, Kozlíčková V, Herbena T, Kovář P, (2005) Ant-induced soil modification and its effect on plant below-ground biomass. *Pedobiologia* 49, 127-137.

Ende CN von (1993) Repeated-measures analysis: growth and other time-dependent measures. In: Scheiner SM, Gurevich J (eds) The Design and Analysis of Ecological Experiments. Oxford University Press, Oxford, pp134–157

Finke, D.L. & Denno R.F. (2003) Intra-guild predation relaxes natural enemy impacts on herbivore populations. *Ecological Entomology*, 28, 67–73.

Finke, D.L. & Denno, R.F. (2004) Predator diversity dampens trophic cascades. *Nature*, 429, 407–410.

Finke, D.L. & Denno, R.F. (2005) Predator diversity and the functioning of ecosystems: the role of intraguild predation in dampening trophic cascades. *Ecology Letters*, 8, 1299–1306.

Frouz J, Holec M., Kalčík J, (2003) The effect of *Lasius niger* (Hymenoptera, Formicidae) ant nest on selected soil chemical properties. *Pedobiologia* 47, 205-212.

Halaj, J.& Wise, D.H. (2001) Terrestrial trophic cascades: How much do they trickle? *American Naturalist,* 157, 262–281.

Hodge, M.A. (1999) The implications of intraguild predation for the role of spiders in biological control. *The Journal of Arachnology*, 27, 351–362.

Kling, G.W., Fry, B. & O'Brien, W.J. (1992) Stable isotopes and planktonic trophic structure in Arctic lakes. *Ecology,* 73, 561–566.

Lane DR, BassiriRad H, (2005) Diminishing effects of ant mounds on soil heterogeneity across a chronosequence of prairie restoration sites. *Pedobiologia* 49, 359-366.

Lang, A. (2003) Intraguild interference and biocontrol effects of generalist predators in a winter wheat field. *Oecologia*, 134,144–153.

Polis, G.A., Myers, C.A. & Holt, R.D. (1989) The ecology and evolution of intraguild predation: potential competitors that eat each other. *Annual Review of Ecology and Systematics*, 20, 297–330.

Ponsard, S. & Arditi, R. (2000) What can stable isotopes ($\delta 15N$ and $\delta 13C$) tell about the food web of soil macroinvertebrates. *Ecology*, 81, 852–864.

Reineking, A., Langel, R. & Schikowski, J. (1993) 15-N, 13-C-on-line measurements with an elemental analyzer (carlo erba, NA 1500), a modified trapping box and a gas isotope mass spectrometer (Finnigan, MAT 251). *Isotopenpraxis*, 29, 169–174.

Reznikova, Z.I., and Panteleeva, S.N. (2001) Interaction of the ants *Myrmica rubra* L. as a predator with springtails (Collembola) as a mass prey. *Doklady Biological Science* 380:475-477.

Riechert, S.E. & Bishop, L. (1990) Prey control by an assemblage of generalist predators: Spiders in a garden test systems. *Ecology*, 71, 1441–1450.

Riechert, S.E. & Lawrence, K. (1997) Test for predation effects of single versus multiple species of generalist predators: spiders and their insect prey. *Entomologia Experimentalis et Applicata*, 84, 147–155.

Rosenheim, J.A., Wilhoit, L.R. & Armet, C.A. (1993) Influence of intraguild predation among generalist insect predators on the suppression of herbivore population. *Oecologia*, 96, 439–449.

Sanders, D. & Platner, C. (2007) Intraguild interactions between spiders and ants and top-down control in a dry grassland. *Oecologia*, 150, 611–624.

Scheu S, (2001) Plants and generalist predators as links between the below-ground and above-ground system. Basic and Applied Ecology 2:3-13.

Schmidt, M.H., Lauer A., Purtauf, T., Thies, C., Schaefer, M. & Tscharntke, T. (2003) Relative importance of predators and parasitoids for cereal aphid control. *Proceedings of the Royal Society of London*, 270, 1905–1909.

Schmitz, O.J., Hämback, P.A., Beckerman, A.P. (2000) Trophic cascades in terrestrial systems: a review of the effects of carnivore removals on plants. *American Naturalist*, 155, 141–153.

Snyder, W.E., Snyder, G.B., Finke, L.F. & Straub, C.S. (2006) Predator biodiversity strengthens herbivore suppression. *Ecology Letters*, 9, 789–796.

Symondson, W.O.C., Sunderland, K.D. & Greenstone, H.M. (2002) Can generalist predators be effective biocontrol agents? *Annual Review of Entomology*, 47, 561–594.

Wada, E., Mizutani, H. & Minagawa, M. (1991) The use of stable isotopes for food web analysis. *Critical Reviews in Food Science and Nutrition*, 30, 361–371.

Wise D.H., Moldenhauer, D.M. & Halaj, J. (2006) Using stable isotopes to reveal shifts in prey consumption by generalist predators. Ecological Applications, 16, 865–876.

Wise, D.H. (1993) Spiders in ecological webs. Cambridge University Press, Cambridge.

Wise, D.H. (2004) Wandering spiders limit densities of a major microbi-detritivore in the forest-floor food web. *Pedobiologia*, 48, 181–188.

Chapter 5

Chapter 5

Habitat structure mediates top-down effects of spiders and ants on herbivores

Dirk Sanders, Herbert Nickel, Thomas Grützner and Christian Platner

Abstract

Differences in structural complexity of habitats have been suggested to modify the extent of top-down forces in terrestrial food webs. In order to test this hypothesis we manipulated densities of generalist invertebrate predators and the complexity of habitat structure in a two-factorial design. We conducted two field experiments in order to study predation effects of ants and spiders and, in particular, of the wasp spider *Argiope bruennichi* on herbivorous arthropods such as grasshoppers, plant- and leafhoppers in a grassland. Predator densities were manipulated by removal in habitats of higher and lower structural diversity, and the effects on herbivore densities were assessed by suction sampling. Habitat structure was changed by cutting the vegetation to half its height and removing leaf litter.

We found a significant negative effect of this assemblage of generalist predators on plant- and leafhoppers, which were 1.6 times more abundant in predator removal plots. This effect was stronger in low-structured (cut) than in uncut vegetation. Densities of the most abundant planthopper *Ribautodelphax pungens* (Delphacidae) were 2.2 times higher in predator removal plots. Furthermore adult planthoppers and leafhoppers responded more strongly than juveniles and epigeic species more strongly than hypergeic species. The presence of predators had a positive effect on plant- and leafhopper species diversity. In a second field experiment we tested the exclusive impact of *A. bruennichi* on its prey, and found that its effect was also significant, although weaker than the effect of the predator assemblage. This effect was stronger in grass-dominated vegetation compared to structurally more complex mixed vegetation of grasses and herbs. We conclude that habitat structure and in particular vegetation height and architectural complexity strongly modify the strength of top-down forces and indirectly affect the diversity of herbivorous arthropods.

Key words. Planthoppers, leafhoppers, grasshoppers, *Argiope bruennichi*, predation, field experiment

Introduction

Top-down forces by invertebrate predators on their herbivorous prey and cascade effects on plants play an important role in structuring communities in terrestrial ecosystems (Schmitz, Hambäck, Beckermann 2000; Halaj & Wise 2001; Shurin, Borer, Seabloom, Anderson, Blanchette et al. 2002). Understanding how top-down effects are mediated by habitat structure may improve our knowledge of predator control on herbivores (Finke & Denno 2006). Strong top-down control in terrestrial ecosystems was mostly demonstrated for simply structured communities (e.g. Finke & Denno 2003; 2004; Schmidt, Lauer, Purtauf, Thies, Schaefer et al. 2003), but increasing structural diversity should modify the strength of interactions within the community (Finke & Denno 2002). Structurally complex habitats can provide refuges for herbivore prey groups resulting in lower predation rates (Crowder & Cooper 1982; Savino & Stein 1989). However, complex vegetation may also promote strong top-down effects by reducing antagonistic interactions among predators (Finke & Denno 2002; Corkum & Cronin 2004). A recent meta-analysis showed that habitat structure plays an important role for the abundance of generalist invertebrate predators (Langellotto & Denno 2004). With increasing habitat structure, most predator guilds reached higher densities and should therefore affect herbivore populations more strongly. However, in food webs with a diverse species assemblage, top-down effects are thought to attenuate (Polis & Strong 1996; Polis 1999; Schmitz et al. 2000). Investigations of the link between habitat structure and top-down control could provide insight into important issues regarding biological control of herbivore insect pests.

The predator guild in our grassland system contains mostly spiders and ants. Most ant species are omnivores, being able to prey on a wide range of other invertebrates (Kajak, Breymeyer, Olechowitz 1972), as well as to take up nutrients from plants indirectly by trophobiosis with phloem-feeding insects (Seifert 1996). Generalist predators such as spiders are able to exert strong top-down control on herbivore populations (Riechert & Bishop 1990; Riechert & Lawrence 1997; Schmitz 1998; Finke & Denno 2003; Cronin, Haynes, Dillemuth 2004) and contribute to the control of pest species in agricultural systems (Symondson, Sunderland, Greenstone 2002; Lang 2003; Schmidt et al. 2003). In our grassland sites in central Europe, *Argiope bruennichi* (Scop) is a dominant spider with regard to biomass and density. This species builds characteristic webs with a vertical zigzag ribbon of silk in the lower stratum of the herb layer and feeds on jumping arthropods like planthoppers, leafhoppers and grasshoppers (Malt 1994).

In a first experiment we tested the effect of an assemblage of generalist invertebrate predators including spider and ant species on herbivores. By manipulating densities of the dominant spider species *Argiope bruennichi*, a second experiment was conducted to test our assumption that this species has the strongest impact on grasshopper and leafhopper populations compared to other predators of the assemblage. Differences in structure were achieved by cutting plants and removing leaf-litter in the first experiment and, in the second experiment, by changing the proportions of grass and herbs.

The aim of this study was to investigate the influence of habitat structure on the extent of top-down control by comparing the effects of predators in simple and structurally more complex vegetation. We hypothesize that top-down effects should attenuate with increasing structural complexity by providing refuges for herbivores.

Material and methods

Both experiments were conducted in 2004 in a chalk grassland on a south-exposed hillside near Witzenhausen (Hesse, Germany, 180 m, 51°22´N, 9°50´E), with grasshoppers (Caelifera), plant- and leafhoppers (Auchenorrhyncha: Fulgoromorpha and Cicadomorpha) and aphids (Aphidina) as the most abundant herbivores. The vegetation was dominated by *Brachypodium pinnatum* (L.) which is one of the most important host grasses of Auchenorrhyncha in central Europe (Nickel 2003). The predator guild includes web-building spiders, wandering spiders and ants. Most abundant taxa were the wandering spider *Zora* spec. (Zoridae) with 10 individuals/m^2, the web-builder *Argiope bruennichi* (Araneidae) with up to six immature and two adult individuals/m^2 and the ant species *Myrmica rubra* L. and *M. sabuleti* (Mei) (Myrmicinae) combined with on average one colony/m^2. Ants of the genus *Lasius*, namely *L. alienus* (För) and *L. niger* (L.) (Formicinae), reached lower densities than *Myrmica* with on average one colony/2m2. Other abundant spider species were *Pisaura mirabilis* (Cl.) (Pisauridae), *Alopecosa trabalis* (Cl.) (Lycosidae), *Pardosa lugubris* (Walck.) (Lycosidae), *Aulonia albimana* (Walck.) (Lycosidae), *Micrommata virescens* (Cl.) (Heteropodidae), *Tibellus oblongus* (Walck.) (Philodromidae), *Clubiona* spec. (Clubionidae), *Enoplognatha ovata* (Cl.) (Theridiidae) and *Tenuiphantes* spec. (Linyphiidae) with a mean density for all species combined of 70 individuals/m^2.

Experiment 1: Exclusion of generalist predators

The basic experimental unit was a 1 m^2 area, enclosed by a 30 cm high plastic fence buried 10 cm deep in the ground and equipped with sticky barriers of vaseline on both sides in order to reduce emigration and immigration of ants and spiders. The plots were left open and had no lid to minimize microclimate effects. This experiment was run from May 13th to July 28th in 2004.

The experiment was carried out in a two-factorial design with the factors "habitat structure" and "predator" resulting in four treatment combinations. Each treatment was replicated three times giving a total of twelve plots. Treatments were randomly assigned to the plots. In order to manipulate habitat structure, the grass was frequently cut with a pair of scissors to a height of 10 cm in six enclosures, which resulted in a simple vegetation structure but maintained a habitat for leafhoppers. Cutting was repeated three times during the whole period in order to achieve a constant grass height in the simple structured enclosures. Leaf litter and cut-off from these plots were removed to gain a simple vegetation structure. The other six plots remained unmanipulated with a vegetation height of 20 cm on average.

The low predator-density treatment was established in six plots by collecting spiders and digging out ant nests. Spider populations and ant colonies that became re-established in the removal plots were removed manually twice a week during the entire length of the experiment. Each plot was searched for spiders and ant colonies by one person for ten minutes and, on average, 4 spiders per plot were removed and released to non-removal plots. This was mainly done in the morning when temperatures were colder and insects were mostly inactive to prevent induced movements of grasshoppers and plant- and leafhoppers through searching activity. The predator treatment was achieved by spider addition to the non-removal plots because the enclosures had a negative effect on spider populations (Sanders & Platner 2006). Ant colonies detected in low predator plots were excavated and replaced by soil cores without ants from outside of the plots. In non-removal plots without ant colonies a supplementary ant colony that was excavated outside the plots was added in order to achieve a comparable ant nest density.

Experiment 2: Exclusion of *Argiope bruennichi*

To test the specific effect of the wasp spider *Argiope bruennichi* on the herbivore guild an additional field experiment was conducted. Twelve plots with an inside area of 1 m^2 delimited by fences of 1 m height were placed in the respective grassland. To evaluate the effect of structural diversity, the experiment was set up in a two-factorial design with the factors "Argiope" and "vegetation structure". Six plots were established in monospecific stands of *Brachypodium pinnatum* and the other six in vegetation composed of *B. pinnatum* and a mixture of different herbs (*Galium mollugo, Achillea millefolium, Lotus corniculatus, Hypericum perforatum, Medicago lupulina, Sanguisorba minor, Agrimonia eupatoria*). The experiment was run for three weeks from August 16th until September 6th 2004. The fence had a mesh size of 5 mm, which allowed leafhoppers and planthoppers to pass through, in contrast to adult *A. bruennichi* and grasshoppers. All plots contained on average 10 grasshopper individuals. In six plots the natural density of two adult female A. bruennichi was established, while in the other six plots *A. bruennichi* was manually removed. Plots were controlled twice a week.

Sampling

At the end of both experiments the invertebrate fauna was sampled using a motor-driven suction sampler (Stihl, Germany; 10 s/sample using a 0.036 m² sampling cylinder). The complete ground area of each enclosure, i.e. one m2, was sampled. Spiders, ants, grasshoppers and plant- and leafhoppers were identified to species level while other arthropods were assigned to higher taxa. In planthoppers and leafhoppers immatures and adults and epigeic and hypergeic species, i.e. those living close to the ground and those living in higher strata (classified after Peter 1981; Nickel 2003) were analysed separately.

Data analysis

The effects of the predator treatment and the habitat structure were tested by an analysis of variance (ANOVA). If necessary, data were log-transformed to meet assumptions of normality and homogeneity of variances. All statistical analyses were performed with SAS (Version 8). We assessed the effect of the predator assemblage and *A. bruennichi* by using the log ratio {ln

(Np+/Np-)} of plant- and leafhopper densities in the presence (Np+) and absence (Np-) of predators (Osenberg, Sarnelle, Cooper 1997; Hedges, Gurevitch, Curtis 1999).

Results

Experiment 1: Generalist predator exclusion

The manipulation of spider and ant densities inside the enclosures proved to be successful. A total of 354 spiders (260 web-builders and 94 wandering spiders) were removed. Densities of the most abundant web-builder *A. bruennichi* were significantly higher in non-removal plots with densities of 4 to 8 individuals/m^2 compared to the removal treatment with less than 1/m^2 (Fig. 1, Table 1). Densities of the most abundant wandering spider *Zora* spec. were 6 times higher in the predator treatment (Fig. 1, Table 1). Colonies of *Lasius alienus* För. and *Myrmica* spp. were frequently removed from removal plots, and this treatment lead to a significant reduction of *Myrmica* colonies (Table 1). Also the density of the most abundant ant (*Myrmica sabuleti*) was higher in non-removal plots (Fig.1, Table 1). For all ant species (*Myrmica*, *Lasius* and *Lepthothorax*) the effect of manipulation was marginally significant (Table 1).

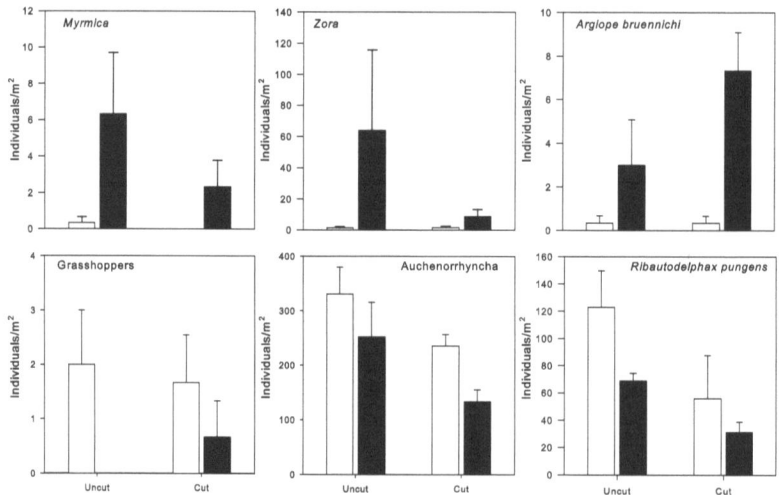

Fig. 1 Mean abundance (+1SE) of *Myrmica*, *Zora*, *Argiope bruennichi*, grasshoppers, Auchenorrhyncha and the most abundant leafhopper species *Ribautodelphax pungens* in different treatments (n = 3). Treatment combinations with normal (Non-removal) and reduced (Removal) ant and spider density in cut (10 cm height) and natural (20 cm height) vegetation.

The density of grasshoppers was affected by the predator treatment only in high vegetation enclosures (ANOVA F1,4 = 10.69, p = 0.031, Fig. 1). It was rather low in 2004 (maximum: 2 ind/m^2), which was probably a result of the cold weather during summer. There was a strong effect of cutting the vegetation on hemipterans. Densities of hemipterans such as planthoppers, leafhoppers, aphids and heteropteran bugs declined from 411 individuals/m^2 in uncut plots to 211 individuals/m^2 in cut plots (Table 1). The presence of ant colonies, *A. bruennichi* and other spiders together had a negative impact on the abundance of planthoppers, leafhoppers and grasshoppers (Fig. 1, Table 1). Planthoppers and leafhoppers were significantly more abundant in the removal-plots than in those with natural predator densities (280 individuals per m^2 compared to 170 individuals per m^2, Table 1).

Among the Auchenorrhyncha the two most abundant species *Ribautodelphax pungens* (Rib.) and *Recilia coronifer* (Marsh.) were most severely affected by the predator treatment (Fig. 1, ANOVA for predator effect: F1,8 = 6.08, p = 0.039; F1,8 = 7.49, p = 0.026, respectively) while for the remaining 29 species no significant effects were observed (see Appendix A). Adult planthoppers and leafhoppers showed a strong response to predator treatment (Fig. 2, ANOVA F1,8 =

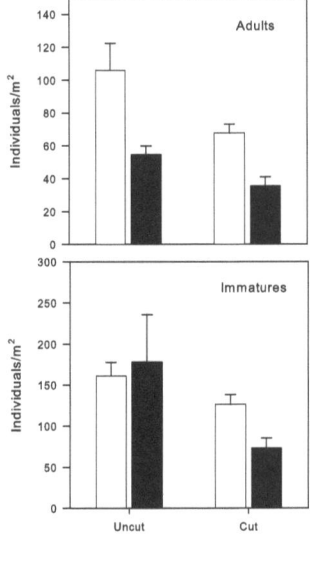

Fig. 2 Mean abundance (+1SE) of juvenile and adult planthoppers and leafhoppers in different treatments (n = 3). Treatment combinations with normal (Non-removal) and reduced (Removal) ant and spider density in cut (10 cm height) and natural (20 cm height) vegetation.

24.25, p = 0.001), while immatures showed no response (Fig. 2, ANOVA F1,8 = 1.54, p = 0.260). Epigeic species were also strongly affected (Fig. 3, ANOVA F1,8 = 14.52, p = 0.005), but no effects could be found for hypergeic species (Fig. 3, ANOVA F1,8 = 1.52, p = 0.282).

Table 1 Response of arthropod abundance to the predator and cutting treatments (Fig. 1). Hemiptera= Planthoppers, leafhoppers, aphids and heteropteran bugs. All comparisons were analyzed using a two-way ANOVA. Data were log-transformed (log$_{10}$X+1). Df for model = 3,8 and treatment = 1,8; *indicates significant treatment effects (p<0.05), bold face= p< 0.06.

Source	Model		Predator		Cutting		Predator*Cutting	
	F	P	F	P	F	P	F	P
Argiope	5.82	0.0207*	13.57	0.0062*	1.94	0.2007	1.94	0.2007
Zora	4.88	0.0324*	11.92	0.0087*	2.08	0.1875	0.65	0.4430
Myrmica	4.11	0.0488*	10.00	0.0133*	0.57	0.4735	1.76	0.2210
Ants	1.84	0.2182	4.95	0.0567(*)	0.57	0.4736	0.00	0.9652
Ant colonies	2.73	0.1137	5.87	0.0417*	1.16	0.3120	1.16	0.3120
Auchenorrhyncha	3.68	0.0624	5.42	0.0483*	5.20	0.0521(*)	0.42	0.5343
Grasshoppers	2.02	0.1897	5.26	0.0510(*)	0,09	0,7663	0,71	0,4242
Hemiptera	3.10	0.0889	2.56	0.1480	6.75	0,0317*	0,00	0,9971

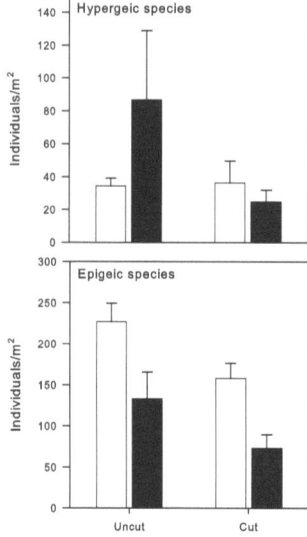

Fig. 3 Mean abundance (+1SE) of planthopper and leafhopper species living in the lower and higher stratum of the herb layer. Treatment combinations with normal (Non-removal) and reduced (Removal) ant and spider density in cut (10 cm height) and uncut (20 cm height) vegetation.

There was a positive effect of the predator treatment on Auchenorrhyncha diversity (Fig. 4). In total we found 21 species in predator removal plots, but 27 in predator plots. Average species richness was not affected by predator treatment (Fig. 4 A; ANOVA $F_{1,8}$ = 1.58, p = 0.244). Values of the Shannon-Wiener index for species diversity were marginally significantly higher for the predator treatment (Fig. 4 B, ANOVA $F_{1,8}$ = 4.83, p = 0.059).

Experiment 2: *Argiope bruennichi* exclusion

Table 2 shows that our manipulation of the *A. bruennichi* density was successful. Each non-removal enclosure contained one or two adult female spiders per m^2 (Fig. 5). This density was commonly recorded by us in the surrounding grassland. Effects of *A. bruennichi* on planthopper and leafhopper densities were strong in pure grass stands (ANOVA for *Argiope* effect: $F_{1,4}$ = 12.57, p = 0.024) but there was no significant effect in the plots with a grass-herb mixture (Fig. 5). Effects on grasshoppers were not significant (Fig. 5, Table 2).

Comparison of predator effects

We calculated the log ratio in order to compare predator effects on plant- and leafhopper densities. The effect was stronger for the predator assemblage in experiment 1 (log ratio –0.46) than for *A. bruennichi* alone in experiment 2 (log ratio –0.11). The effect of ants and spiders in experiment 1 was stronger in cut (log ratio –0.60) than in uncut plots (log ratio –0.33). Prey suppression through *A. bruennichi* in experiment 2 was stronger in pure grass plots (log ratio –0.23) than in herb-grass-mixture plots (log ratio –0.07).

Table 2 Response of arthropod abundance to *Argiope* removal and habitat structure. All comparisons were analyzed using a two-way ANOVA. Data were log-transformed (log10X+1). Df for model = 3, 8 and treatment = 1, 8; *indicates significant treatment effects (p<0.05).

Source	Model		Argiope		Structure		Argiope*Structure	
	F	P	F	P	F	P	F	P
Argiope	42.25	<.0001*	84.51	<.0001*	0.00	1.0000	0.00	1.0000
Grasshoppers	1.04	0.4263	1.30	0.2872	0.21	0.6584	1.60	0.2405
Auchenorrhyncha	22.99	0.0003*	3.52	0.0974	62.37	<.0001*	3.07	0.1178

Discussion

We found that simply structured habitats enhanced plant- and leafhopper suppression by spiders and ants and by *Argiope bruennichi*. This result can be explained by a change in the composition of the predator assemblage with changing habitat structure in the first experiment and by a higher number of refuges for prey in more complex habitats. Distribution of predator groups in cut and uncut vegetation was different in non-removal plots. *A. bruennichi* tended to be more abundant in cut than in uncut plots (Fig. 1). Thus, the stronger effect size on Auchenorrhyncha populations in the simple-structured habitats may be caused by higher densities of *A. bruennichi*, which was also able to affect plant- and leafhopper densities strongly on its own in the second experiment. Additionally, prey abundance was also strongly affected by cutting the vegetation: There was a significant negative effect on the population size of all Hemiptera, including aphids, heteropteran bugs, planthoppers and leafhoppers. Additionally, habitat structure also directly influences the strength of top-down effects. We assume that prey availability was better in cut areas, where the encounter frequency between prey and predators is higher. A more complex habitat structure can provide shelter from predators and reduce effects of predation (McNett & Rypstra 2000, Finke & Denno 2002), which is also supported by our results. We conclude that changes in habitat structure strongly affect both herbivores and predators, and that differences in the strength of top-down effects may be mediated by the species identity of predators and the density of the prey. The structure itself effects top down forces strongly by providing refuge from predation.

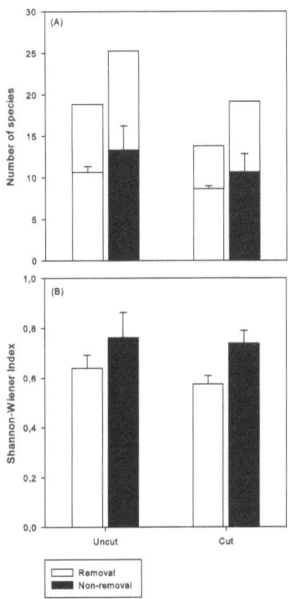

Fig. 4 (A) total and mean species number (+1SE) of Auchenorrhyncha in different treatments of ants and spiders exclusion experiment and (B) Shannon Wiener index (+1SE) for species richness of Auchenorrhyncha.

For immature planthoppers and leafhoppers no effect of predator treatment was found whereas adults responded strongly (Fig. 2). Adults have a higher agility than earlier instars due to their longer legs (Wilson & McPherson 1981; Ballou, Tsai, Wilson 1987) and better jumping capability (pers. observation). Nymphs are more sedentary on the lower parts of plants while mating, oviposition and dispersal are behavioral characteristics of adult plant- and leafhoppers and involve a higher amount of movements with an increased probability of predator-prey encounters.

Only those species of Auchenorrhyncha, that were known to live close to the soil, were strongly affected by the predator treatment, whereas those living in higher strata showed no response, probably due to higher predation rates in the lower strata. The encounter frequency between predators and prey may be higher in the lower strata, because most predators such as ants and wandering spiders live closer to the litter layer. Since webs of *A. bruennichi* were lower in cut vegetation, this may also affect epigeic species more strongly.

The assumption that *A. bruennichi* exerted strong effects on Auchenorrhyncha could be confirmed by our second experiment. The effect of *A. bruennichi* in pure grass plots was similar to the effect size of the predator assemblage in uncut plots. Similarly to the impact of the assemblage, *A. bruennichi* affected herbivore populations much stronger in simply structured pure grass stands than in an architectural more complex grass-herb mixture. There was a strong effect of vegetation on total Auchenorrhyncha abundance. A high percentage of herbs caused a decrease in Auchenorrhyncha abundance because most of them were grass-feeding species. But this does not explain the absence of effects by *A. bruennichi* in enclosures with grass-herb mixture. Many herbs show more horizontal architectural components, notably leaves, which reduce the availability of space for large orb-webs. In this experiment, habitat structure appeared to influence top-down effects of *A. bruennichi* by changing the effectiveness of their webs.

Total number of Auchenorrhyncha species and Shannon-Wiener index were positively correlated with predator presence (Fig. 4). Such an effect of predators on the diversity of their prey has been demonstrated for aquatic systems (Paine 1966, 1971, Werner 1991) and for grassland plants (Lubchenco 1978, Schmitz 2003). In all these studies the abundance of dominant species

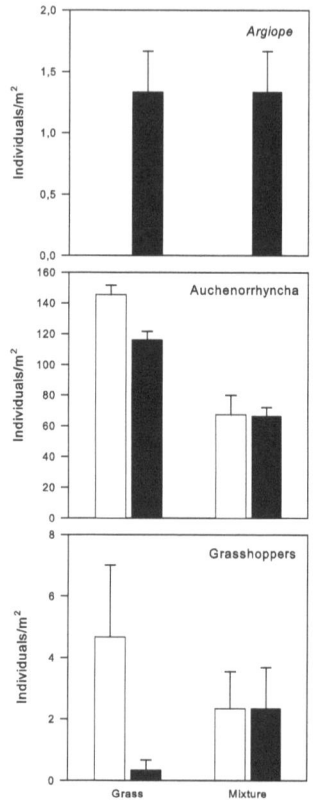

Fig. 5 Mean abundance (+1SE) of *Argiope bruennichi* and their prey groups in the *Argiope*-exclusion experiment. Treatments with normal *Argiope* density and *Argiope* removed in grass-only and grass-herb-mixed vegetation.

was reduced and competitively inferior species were able to invade. One possible explanation for this effect in our experiment is release from competitive pressure from the two dominant leafhoppers species *R. pungens* and *R. coronifer*, which had lower densities in the presence of predators. *R. pungens* is monophagous on the dominant grass *B. pinnatum* (Nickel 2003) and resources of this grass may be better available for this species than for others. Among sap feeders competitive interactions tended to be frequent more often compared to other herbivores (Denno, Mclure & Ott 1995). There is yet no evidence for such a competitive relationship between leafhopper species in our system, however, in an other study predator removal increased herbivore densities and produced evidence of increased interspecific competition (Edson 1985).

Increasing habitat complexity strongly affected the extent of top-down control by spiders and ants probably by providing spatial refuges and by changing the predator assemblage. Effects on planthoppers and leafhoppers were stronger in less complex vegetation. We conclude that *A. bruennichi* is an important predator in this interaction, because there was also a strong effect on Auchenorrhyncha densities in the second experiment with removal of only *A. bruennichi*. Therefore this species, which has only relatively recently invaded most of Central Europe, has gained the role of a keystone predator, which can severely affect herbivore populations. In our experiments the structure of the habitat strongly modified the extent of top-down forces and indirectly affected the diversity of Auchenorrhyncha species.

Acknowledgments

We thank Matthias Schaefer, Peter Hambäck, Jason Tylianakis, Sharon Cooper and four anonymous reviewers for comments and suggestions that improved this manuscript.

References

Ballou, J. K., Tsai, J. H., Wilson, S. W. (1987). Delphacid planthoppers Sogatella kolophon and Delphacodes idonea (Homoptera: Delphacidae): Descriptions of immature stages and notes on biology. Annals of the Entomological Society of America, 80, 312-19.

Corkum, L. D., & Cronin, D. J. (2004). Habitat complexity reduces aggression and enhances consumption in crayfish. Journal of Ethology, 22, 23-27.

Cronin, J. T., Haynes, K. J., & Dillemuth, F. (2004). Spider effects on planthopper mortality, dispersal, and spatial popular dynamics. Ecology, 85, 2134-2143.

Crowder, L. B., & Cooper, W. E. (1982). Habitat structural complexity and the interaction between bluegills and their prey. Ecology, 63, 1802-1813.

Denno, R.F., McClure, M.S., Ott, J.R. (1995). Interspecific interactions in phytophagous insects: Competition reexamined and resurrected. Annual Review of Entomology 40, 297-331.

Edson, JL. (1985). The influences of predation and resource subdivision on the coexistence of goldenrod aphids. Ecology, 66,1736-43.

Finke, D.L., & Denno R.F. (2002). Intraguild predation diminished in complex-structured vegetation: Implications for prey suppression. Ecology, 83, 643-652.

Finke, D.L., & Denno, R.F. (2003). Intra-guild predation relaxes natural enemy impacts on herbivore populations. Ecological Entomology, 28, 67-73.

Finke, D.L., & Denno, R.F. (2004). Predator diversity dampens trophic cascades. Nature, 429, 407-410.

Finke, D.L., & Denno, R.F. (2006). Spatial refuge from intraguild predation: implications for prey suppression and trophic cascades. Oecologia, 149, 265-275.

Halaj J., & Wise D.H. (2001). Terrestrial trophic cascades: How much do they trickle? American Naturalist, 157, 262-281.

Hedges, L.V., Gurevitch, J., & Curtis, P.S. (1999). The meta-analysis of response ratios in experimental ecology. Ecology, 80, 1150-1156.

Kajak, A., Breymeyer, A., Pêtal, J., & Olechowicz, E. (1972). The influence of ants on the meadow invertebrates. Ekologia Polska, 20 (17), 163-171.

Lang, A. (2003). Intraguild interference and biocontrol effects of generalist predators in a winter wheat field. Oecologia 134,144-153.

Langellotto, G.A., & Denno, R.F. (2004). Response of invertebrate natural enemies to complex-structured habitats: a meta-analytical synthesis. Oecologia 139,1-10.

Lubchenco, J. (1978). Plant Species Diversity in a Marine Intertidal Community: Importance of Herbivore Food Preference and Algal Competitive Abilities. American Naturalist, 112, No. 983, 23-39.

Malt, S. (1994). Trophische Beziehungen ausgewählter netzbauender Araneen in Halbtrockenrasen unter besonderer Berücksichtigung von *Argiope bruennichi* (Scopoli 1772) - Mitteilungen der Deutschen Gesellschaft für Allgemeine und Angewandte Entomologie, 435-446.

McNett, B.J., & Rypstra A.L. (2000). Habitat selection in a large orb-weaving spider: vegetational complexity determines site selection and distribution. Ecological Entomology 25, 423-432.

Nickel, H. (2003). The leafhoppers and planthoppers of Germany (Hemiptera, Auchenorrhyncha): patterns and strategies in a highly diverse group of phytophagous insects. Sofia–Moscow: Pensoft and Keltern: Goecke & Evers.

Osenberg, C.W., Sarnelle, O., & Cooper, S.D. (1997). Effect size in ecological experiments: the application of biological models in meta-analysis. American Naturalist, 150, 798-812.

Paine, R.T. (1966). Food Web Complexity and Species Diversity. American Naturalist, 100, 65-75.

Paine, R.T. (1971). A Short-Term Experimental Investigation of Resource Partitioning in a New Zealand Rocky Intertidal Habitat. Ecology, 52(6), 1096-1106.

Peter, H.U. (1981). Further investigations on the settling of niches by leafhoppers (Homoptera Auchenorrhyncha) on the meadows of the Leutra Valley near Jena. Zoologisches Jahrbuch Systematik, 108, 563-588.

Polis G.A. (1999). Why are parts of the world green? Multiple factors control productivity and the distribution of biomass. Oikos, 86, 3-15.

Polis G.A., & Strong D.R. (1996). Food web complexity and community dynamics. American Naturalist, 147, 813-846.

Riechert, S.E., & Bishop, L. (1990). Prey control by an assemblage of generalist predators: spiders in garden test systems. Ecology, 71, 1441-1450.

Riechert, S.E., & Lawrence K. (1997). Test for predation effects of single versus multiple species of generalist predators: spiders and their insect prey. Entomological Experimental Applications, 84, 147-155.

Sanders, D., & Platner, C. (2006). Intraguild interactions between spiders and ants and top-down control in a dry grassland. Oecologia. DOI 10.1007/s00442-006-0538-5.

Savino, J.F., & Stein, R.A. (1989). Behavioral interactions between fish predators and their prey: Effects of plant density. Animal Behaviour, 37, 311-321.

Schmidt M.H., Lauer A., Purtauf T., Thies C., Schaefer M., & Tscharntke T. (2003). Relative importance of predators and parasitoids for cereal aphid control. Proceedings of the Royal Society of London, Series B: Biological Sciences, 270, 1905-1909.

Schmitz, O.J. (1998). Direct and indirect effects of predation and predation risk in old-field interaction webs. American Naturalist, 151, 327-342.

Schmitz, O.J. (2003). Top predator control of plant biodiversity and productivity in an old-field ecosystem. Ecology Letters, 6, 156–163.

Schmitz, O.J., Hämback, P.A., & Beckerman, A.P. (2000). Trophic cascades in terrestrial systems: a review of the effects of carnivore removals on plants. American Naturalist, 155, 141–153.

Seifert, B. (1996). Ameisen: beobachten, bestimmen. Augsburg : Naturbuch-Verlag.

Shurin, J.B., Borer, E.T., Seabloom, E.W., Anderson, K., Blanchette, C.A., Broitman, B., Cooper, S.D., & Halpern, B. (2002). A cross-ecosystem comparison of the strength of trophic cascades. Ecology Letters, 5, 785-791.

Symondson, W.O.C., Sunderland K.D., & Greenstone H.M. (2002). Can generalist predators be effective biocontrol agents? Annual Review of Entomology, 47, 561-594.

Werner, E.E. (1991). Nonlethal effects of a predator on competitive interactions between two anuran larvae. Ecology, 72, 1709-1720.

Wilson S. W., McPherson, J. E. (1981). Life histories of *Anormenis septentrionalis*, *Metcalfa pruinosa*, and *Ormenoides venusta* with descriptions of immature stages. Annals of the Entomological Society of America,74, 299–311.

Chapter 6

Chapter 6

Small scale habitat fragmentation affects generalist predator diversity: Implications for top-down control of spiders

Chapter 6

Abstract

Habitat fragmentation is a common phenomenon in the cultural landscape. We studied the effects of small-scale fragmentation on the species diversity of the most common generalist predators such as web-building spiders and hunting spiders in grasslands and their impact on prey arthropods. Grassland fragments of different size and varying distance to the surrounding habitat were created by frequent mowing. We established these experiments in seven grassland habitats ranging from more natural fallows to meadows in Central Germany over a period of two years. The response of spider density and diversity and the density of other arthropods were assessed by suction sampling and heat extraction from soil samples.

Spiders responded to the treatment effect of fragment size, however the varying distance of the fragments to surrounding habitat had no effect on the density and diversity of the spider assemblages. In the first year diversity of spiders was higher in smaller fragments compared to large fragments. For hunting spiders average species number was highest in small fragments while for web-building spiders medium fragments contained most species. In the second year the pattern was the same for hunting spiders and their density was also significantly higher in small fragments. The pattern for web-building spiders was slightly different compared to year before. Average species number of web-builders was highest in small fragments and lowest in medium fragments. Collembola density was reduced by 30% on small fragments indicating a stronger top-down effect of a more diverse hunting spiders assemblage in the first year. Biomass of plants did not respond significantly to fragmentation, however, biomass tended to be lower in larger fragments, where density and diversity of spiders was low.

We conclude that spiders, especially hunting spiders, were more diverse on small grassland fragments compared to larger fragments and the surrounding habitat. Possible prey groups such as collembolans were present with lower densities in small fragments indicating a stronger top-down control of a higher diverse predator guild.

Key words: web-building spiders, hunting spiders, Collembola, field experiment

Introduction

The rapid modification of the landscape by humans over the past half century has led to an increased fragmentation of natural habitats. Loss and isolation of near-natural habitats are extremely widespread and pose perhaps the most serious threat to biological diversity (Saunders et al. 1991, Collinge 2000, Simberloff 2000). Disruptions to continuous habitats are known to alter many ecological processes and interactions among species (Kruess and Tscharnke 1994, Groppe et al. 2001, Goverde et al. 2002). Theory predicts that higher trophic levels are more prone to extinction caused by fragmentation due to their smaller population size and dependence on prey populations (Holt 1996, Duffy 2002, 2003). A change in predator abundance and diversity may change the strength of trophic links in food webs as demonstrated in experiments by Finke and Denno (2004, 2005) and Snyder et al. (2006). Braschler et al. (2003) found that habitat fragmentation caused an increase in aphid density due to higher ant-attending rates in these fragments, which probably protected the aphids from other predators. Thus the dominant impacts of biodiversity change on ecosystem functioning appear to be trophically mediated (Duffey 2003). Spiders are the most abundant generalist predators in grasslands which can strongly affect arthropod populations (Riechert and Bishop 1990, Wise 1993, Riechert & Lawrence 1997). According to their different strategies in catching prey spiders can be divided into two main functional groups: web-building spiders and hunting spiders. Web-building spiders are dependent on a suitable habitat structure for building their webs, while hunting spiders are more sensible to microclimate conditions of the habitat. Spiders are able to colonize habitats during their juvenile life stages by ballooning and especially many adult hunting spiders are very mobile.

We examined the effect of experimental small-scale habitat fragmentation on the diversity and density of spiders. We hypothesise that fragment size and the distance of fragments to the natural habitat strongly influence the species composition of the generalist predator guild. This differences in species diversity should also affect important prey groups.

Material and methods

Study sites

The experiment was conducted in seven different grassland sites in Central Germany (Lower Saxony, Hesse, Thuringia). Two experimental sites established in dry grasslands with *Brachipodium pinnatum* (L.) as dominant grass species and two sites on damp grasslands with mainly *Carex* species. Additionally we chose two meadows and one fallow dominated by *Arrhenatherum elatius*. Hence, these seven sites include the gradients of land use and of soil moisture.

Experiment

The fragmentation of the grasslands was created by mowing of the vegetation around the fragments using a scythe mower with a mowing bar width of 72 cm. The experiment ran from April 2005 until September 2006 and mowing of the vegetation started in April 2005 and was repeated four up to five times during each vegetation period in 2005 and 2006 to maintain habitat fragmentation. After mowing, the cut vegetation was removed from the experimental area. One experimental unit, called block, contained three large (7.2 m^2), three medium (1,8 m^2) and three small (0,45 m^2) fragments of grassland in three different distances (72, 144 and 288 cm) to the surrounding vegetation (Fig. 1). This block was replicated in seven grassland sites in central Germany. Within each block, the position of the fragments with different size and distance to unmown grassland was varied.

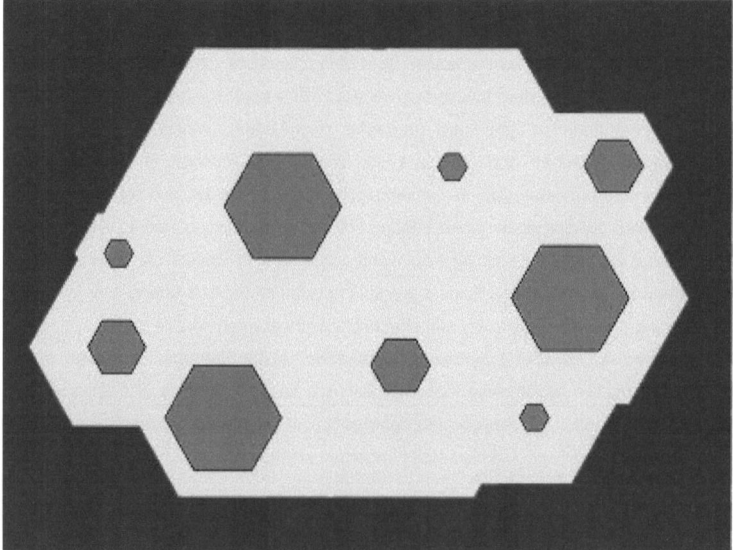

Fig. 1 Design of one block of the experiment. A block contained three large (7.2 m^2), three medium (1,8 m^2) and three small (0,45 m^2) fragments of grassland in three different distances (72 cm, 144 cm and 288 cm) to the surrounding vegetation, their position in each block was varied. The frequently mown area is shown in light grey.

Sampling

The fauna was sampled with a suction sampler (Stihl SH 85, Germany; 10 s/sample using a 0.036 m² sampling cylinder) in June and September in both years. Additionally in September 2006, we took soil samples from 0.036 m² soil cores, which subsequently were treated by heat extraction (Kempson 1963, Schauermann 1982). Both sampling methods were used for the same area; after placing the sampling cylinder on the ground and taking the suction sample, a soil core of 7 cm thickness was taken from the same area and transferred to the laboratory for heat extraction. As result we had a suction sample and a soil sample from the same surface. Arthropods which were not caught by suction sampling were consequently found in the samples from soil sample, which is especially important for estimating real density of ground living spiders. To account for possible edge effects we additionally took suction samples from the margin and the centre of the large fragments. Spiders were identified to species level, while other arthropods were assigned to higher-ranking taxa. In order to estimate plant biomass, plants that were sampled in the 0.036 m² soil cores were dried for 72 h at a temperature of 60°C, and the dry weight was measured.

Statistics

To test the effect of fragment size, distance to the natural habitat and the block effect we used a analysis of variance (ANOVA procedure, SAS version 8). All abundance and biomass data were log-transformed to meet assumptions of normality and homogeneity of variances.

Results

Spiders responded to the treatment effect of fragment size, however the varying distance of the fragments to surrounding habitat had no effect on the density and diversity of the spider assemblages (Table 1).

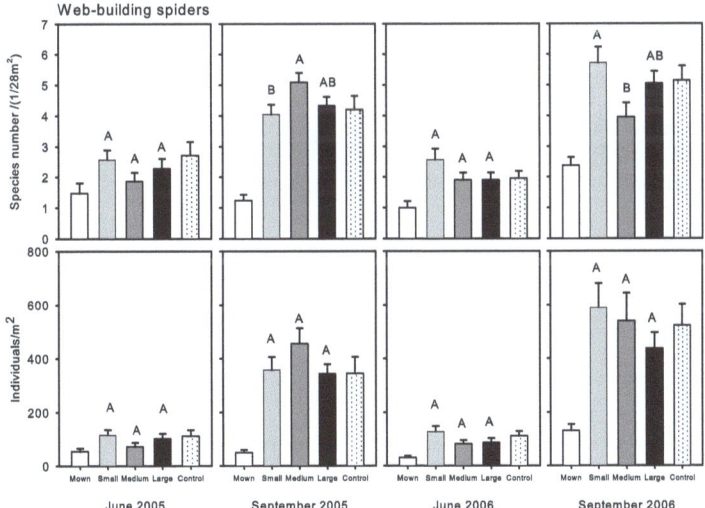

Fig. 2 Mean species diversity (per 0.036 m^2) and mean abundance of web-building spiders at the mown area (matrix), the fragments of different size (small 0,45 m^2, medium 1,8 m^2 and large 7.2 m^2) in the fragmentation experiment and in control samples in undisturbed vegetation. For statistical analyses see table 1.

Table 1 Response of spider diversity and density to "Patch size" and "Distance" in the fragmentation experiment. Data from suction samples for spring, and combined samples from soil cores and suction sampling for summer. All comparisons were analyzed using a two-way ANOVA. Data were log-transformed (log10X+1). Df for model = 14;48, treatment = 2;48 and block 6;48 ; bold numbers indicate significant treatment effects (p<0.05).

Source	Model		Patch size (P)		Distance (D)		P × D		Block	
	F	P	F	P	F	P	F	P	F	P
Spring 2005										
Web-builders										
Species diversity	1.56	0.1265	1.58	0.2162	1.13	0.3308	0.26	0.8997	2.56	**0.0313**
Density	1.29	0.2466	2.08	0.1360	0.17	0.8455	0.13	0.9718	2.18	0.0609
Hunting spiders										
Species diversity	3.24	**0.0012**	0.10	0.9035	1.29	0.2854	0.64	0.6376	6.66	**<.0001**
Density	2.07	**0.0317**	0.72	0.4931	0.63	0.5345	0.80	0.5330	3.84	**0.0033**
Summer 2005										
Web-builders										
Species diversity	1.48	0.1555	3.42	**0.0409**	0.70	0.5034	1.40	0.2492	1.15	0.3493
Density	1.99	0.0398	1.62	0.2080	0.18	0.8367	0.25	0.9056	3.86	**0.0032**
Hunting spiders										
Species diversity	2.74	0.0048	3.98	**0.0251**	0.54	0.5837	0.44	0.7761	4.58	**0.0010**
Density	2.03	0.0353	0.70	0.4996	0.93	0.4029	0.48	0.7536	3.87	**0.0031**
Spring 2006										
Web-builders										
Species diversity	0.87	0.5897	1.21	0.3029	0.18	0.8394	1.06	0.3884	0.87	0.5220
Web-builders Density	1.05	0.4265	1.60	0.2125	1.13	0.3327	0.22	0.9244	1.39	0.2397
Hunting spiders										
Species diversity	3.37	**0.0008**	3.23	**0.0483**	1.22	0.3053	0.58	0.6798	5.99	**<.0001**
Density	3.24	**0.0012**	2.23	0.1184	0.74	0.4840	0.79	0.5398	6.05	**<.0001**
Summer 2006										
Web-builders										
Species diversity	5.53	**<0.0001**	6.96	**0.0022**	0.98	0.3838	0.76	0.5580	9.75	**<0.0001**
Density	8.36	**<0.0001**	0.43	0.6505	0.25	0.7767	0.93	0.4561	18.65	**<0.0001**
Hunting spiders										
Species diversity	3.37	**0.0008**	2.22	0.4225	0.88	0.4225	2.38	0.0646	5.25	**0.0003**
Density	4.55	**<0.0001**	5.50	**0.0071**	1.25	0.2958	1.93	0.1199	7.08	**<0.0001**

Web-building spiders

Web-building spiders were generally more diverse with up to 6 species per 0.036 m^2 than hunting spiders (4 species per 0.036 m^2) and also numerically the dominant spider group with densities of 110 in June and 500 individuals/m^2 in September. The first sampling two months after the start of fragmentation revealed no effects of fragment size and distance on species diversity and density of spiders (Table 1). However, four months later in September, web-building spiders were significantly more diverse in medium fragments with an average species number of five compared to small fragments with four species per 0.036 m^2, whereas large fragments had an intermediate position (Fig. 2). The same pattern could be observed for the density of web-builders, although this was not significant. Similarly to 2005 in the second year in June no effects were found. In contrast to the previous year in September species diversity was higher in small fragments and lowest in medium fragments (Fig. 2, Table 1).

Hunting spiders

Hunting spiders had lower densities than web-builders with on average 50 individuals/m^2 in June and 120 individuals/m^2 in September. They seem to profit from fragmentation, due to their increased density in smaller patches compared to the control. Hunting spiders were most diverse in small fragments in September 2005 and their species diversity declined with increasing patch size (Table 1, Fig. 3), however, no effect on hunting spider density was found at this occasion. In the second year, June samples confirmed the pattern of the summer for hunting spider diversity. In September 2006 also the density of hunting spiders was two times higher in small fragments compared to large fragments, which had similar densities to the control (Fig. 3, Table 1).

From large fragments samples were taken from the centre and the margin of the fragment to evaluate possible edge effects, however, no edge effect on spider density and species diversity was found (ANOVA test of edge effect on density $F_{1;61}$ =0.01; p = 0.9349 and diversity $F_{1;61}$ = 0.16; p = 0.6915 of spiders).

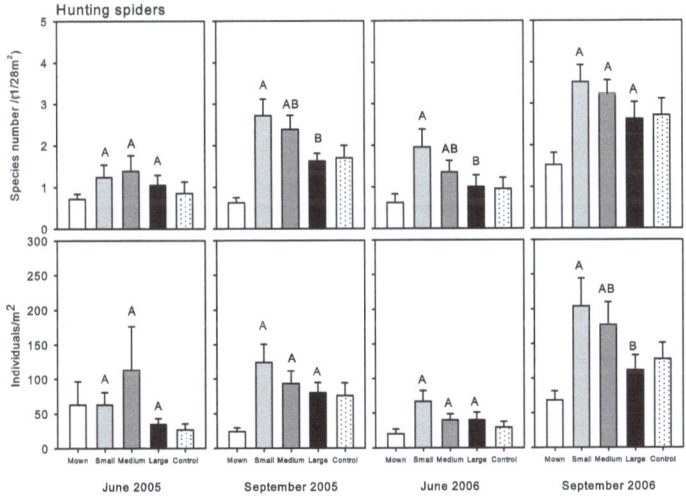

Fig. 3 Mean species diversity (per 0.036 m^2) and mean abundance of hunting spiders at the mown area (matrix), the fragments of different size (small 0,45 m^2, medium 1,8 m^2 and large 7.2 m^2) in the fragmentation experiment and in control samples in undisturbed vegetation. For statistical analyses see table 1.

Prey groups

Epigeic Collembola as most abundant detritivores and Auchenorrhyncha as the dominant herbivorous group were also tested for treatment effects. Members of the Auchenorrhyncha did not respond to treatment (Table 2), while the density of Collembola showed a pattern contrary to that of hunting spiders density and diversity and were reduced by 30% in population size (Fig. 4, Table 2). The water content of the soil, which might affect Collembola density, did not differ between the fragments of different sizes (ANOVA $F_{2;60} = 0.02$; $p = 0.9797$). Biomass of herbs tended to be higher in the smaller fragment compared to large fragments (Fig. 5, ANOVA $F_{2;48} = 1.47$; $p = 0.2411$).

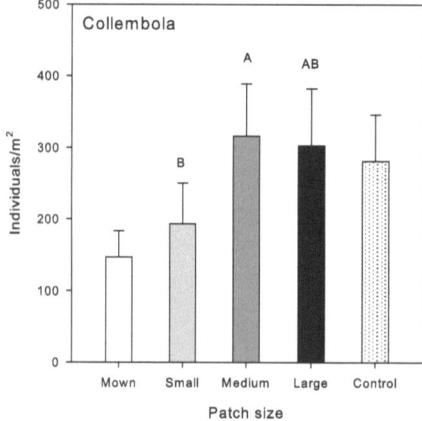

Fig. 4 Mean abundance of Collembola (spingtails) at the mown area (matrix) and fragments of different size (small 0,45 m^2, medium 1,8 m^2 and large 7.2 m^2) and in control samples in undisturbed vegetation. For statistical analyses see table 2.

Table 2 Response of Collembola and Auchenorrhyncha to "Patch size" and "Distance" in the fragmentation experiment. Data from suction samples for spring, and combined samples from soil cores and suction sampling for summer. All comparisons were analyzed using a two-way ANOVA. Data were log-transformed (log10X+1). Df for model = 14;48, treatment = 2;48 and block 6;48 ; bold numbers indicate significant treatment effects (p<0.05).

Source	Model		Patch size (P)		Distance (D)		P × D		Block	
	F	P	F	P	F	P	F	P	F	P
Spring 2005										
Collembola	7,35	<,0001	0,20	0,8162	0,79	0,4584	0,88	0,4823	16,22	<,0001
Auchenorrhyncha	5,88	<,0001	1,53	0,2279	0,24	0,7862	1,10	0,3693	12,39	<,0001
Summer 2005										
Collembola	5,64	<,0001	3,87	**0,0276**	4,08	**0,0230**	1,10	0,4130	9,48	<,0001
Auchenorrhyncha	2,22	0,0204	1,04	0,3600	0,33	0,7235	0,48	0,7519	4,42	**0,0012**
Spring 2006										
Collembola	3,84	0,0002	1,60	0,2120	1,61	0,2103	1,22	0,3166	7,08	<,0001
Auchenorrhyncha	6,87	<,0001	0,44	0,6491	2,25	0,1163	2,51	0,0541	13,45	<,0001
Summer 2006										
Collembola	17,83	<,0001	2,28	0,1137	0,68	0,5128	0,74	0,5689	40,13	<,0001
Auchenorrhyncha	11,99	<,0001	0,43	0,6550	0,75	0,4771	1,11	0,3606	26,83	<,0001

Discussion

Both spider groups, i.e. hunting spiders and web-building spiders, were affected by the small scale fragmentation at the seven grassland habitats. Spiders responded to the treatment effect of fragment size, however the varying distance of the fragments to surrounding habitat had no effect on the density and diversity of the spider assemblages. For web-builder diversity, effects were only strong in September in both years, because most spiders were juvenile in spring and could not be identified at species level. The effect of fragment size on species diversity was different for both years, while web-builder diversity was highest in medium fragments in 2005, a

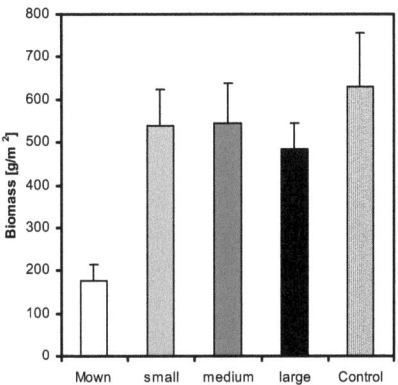

Fig. 5 Biomass of plants at the mown area (matrix) and fragments of different size (small 0,45 m^2, medium 1,8 m^2 and large 7.2 m^2) and in control samples in undisturbed vegetation.

year later diversity was lowest on the same fragment size. Large fragments contained the lowest species diversity in both years, which was similar for hunting spider diversity. Hunting spiders were generally less diverse than web-builders, however their response to fragment size was also strong with higher species numbers on small fragments. For hunting spiders the density was affected in the same way as their diversity. Cronin et al. (2004) found that patch size of cordgrass affected the distribution of spiders in the same way as in our experiment. Density of web-builders and cursorial spiders declined with increasing patch size, whereas effects on diversity were not tested. In an old-field study in Kansas web-building spiders were also more abundant in fragments, particularly along edges, where they can profit from the aerial "drift" of insects from the surrounding habitat (Jackson et al. unpublished data in Debinski and Holt 2000), a pattern that was also found for ground living spiders in an urban fragmentation study (Bolger et al. 2000). While for web-building spiders diversity on the fragments was not higher compared to the surrounding habitat, hunting spider diversity was significantly higher in small patches. Thus, fragmentation increased hunting spider diversity probably by providing more suitable habitats due to a change of the microclimate. Following from the theory of island biogeography (Mac-Arthur and Wilson 1967) smaller, more isolated fragments are expected to retain fewer species than larger fragments. This pattern of declining diversity in smaller patches was actually found in some studies on habitat fragmentation (Baur and Erhard 1995, Laurance and Bierregaard 1996, Collinge 2000). An alternative hypothesis is that species move from the matrix to the remaining habitat patches after a disturbance, e.g. the start of fragmentation, and cause a "crowding" effect (Fahrig and Paloheimo 1988). This may also include a "retreat effect" for mobile taxa, which may use the mown area surrounding the fragments as foraging area (Zschokke et al. 2000). Increasing the amount of edge can subject a habitat to more extreme abiotic influences such as wind and temperature (Saunders et al. 1991). We assumed a positive edge effect on spider diversity in the small fragments, however, the test for differences in species diversity in the margin compared to the centre of larger fragments raised no evidence for such an effect. To summarize, we assume that a "retreat effect" may explain the higher abundance and diversity of hunting spiders in small fragments. For web-builders response as a pure edge effect was unlikely, however, this may be explained by a higher food availability based on the aerial "drift" of insects from the surrounding habitat.

Collembola as one possible prey group responded also to the treatment of fragment size, their density was negatively affected by small fragments. As possible explanation, we could exclude differences in soil moisture due to a less denser vegetation. We assume that the higher density and diversity of spiders, especially hunters, caused the decline in populations size. Collembola are known to be one of the most important prey groups for spiders as generalist predators (Nentwig 1986, Nyffeler 1999, Lawrence and Wise 2000; 2004, Wise 2004) particularly for juveniles (Sanders and Platner 2007). Biomass of plants was only marginally influenced and showed a tendency to lower biomass in large fragments, which would also fit to a stronger top-down control of generalist predators in the smaller fragments. The most abundant herbivorous group in the grassland sites were planthoppers and leafhoppers (Auchenorrhyncha), and in the study of Cronin *et al.* (2004) the density of planthoppers was negatively correlated with spider density. In our experiments Auchenorrhyncha density was not affected.

Fragmentation had a strong influence on spider diversity and density, which generally declined with increasing fragment size. Spiders seem to profit from small fragments and were able to cause a decline in Collembola population size due to a stronger top down control. The distance of the fragments to the natural habitat had no effect on spiders, probably distances in our experiment were too small to cause higher extinction rates.

Acknowledgements

We would like to thank Heiko Bischoff, Simone König, Christina Taraschewski, Herbert Nickel, and Sharon Cooper for their help in the field and Sebastian Schuch, Christel Fischer for providing assistance concerning the huge amount of samples. This study was financially supported by the Deutsche Forschungsgemeinschaft.

References

Baur, B. and Erhardt., A. (1995) Habitat fragmentation and habitat alterations: principal threats to most animal and plant species. GAIA 4:221-226.

Braschler, B., Lampel, G., Baur, B. (2003) Experimental small-scale grassland fragmentation altern aphid population dynamics. Oikos 100:581-591.

Collinge, S.K. (2000) Effects of grassland fragmentation on insect species loss, colonization, and movement patterns. Ecology 81:2211-2226.

Cronin, J.T., Haynes K.J., Dillemuth, F. (2004) Spider effects on planthopper mortality, dispersal, and spatial population dynamics. Ecology 85: 2134-2134

Debinski, D.M. and Holt R.D. (2000) A survey and overview of habitat fragmentation experiments. Conservation biology 14: 342-355.

Bolger, D.T., Suarez, A.V., Crooks, K.R., Morrison, S.A., Case, T.J. (2000) Arthropods in Urban Habitat Fragments in Southern California: Area, Age, and Edge Effects. Ecological Applications 10,1230-1248.

Duffy, J.E. (2002) Biodiversity and ecosystem function: the consumes connection. Oikos, 99, 201–219.

Duffy, J.E. (2003) Biodiversity loss, trophic skew and ecosystem functioning. Ecology Letters, 6, 680–687.

Fahrig, L. and Paloheimo J. (1988) Determinants of local population size in patchy habitats. Theoretical Population Biology 34:194-213.

Finke, D.L. and Denno, R.F. (2004) Predator diversity dampens trophic cascades. Nature, 429, 407–410.

Finke, D.L. and Denno, R.F. (2005) Predator diversity and the functioning of ecosystems: the role of intraguild predation in dampening trophic cascades. Ecology Letters, 8, 1299–1306.

Goverde, M., Baur, B., Erhardt, A. (2002) Small-scale habitat fragmentation effects on pollinator behavior: experimental evidence from the bumblebee *Bombus veteranus* on calcareous grassland. Biological Conservation 104, 293 299.

Groppe K., Steinger T., Schmid B., Baur B., Boller T. (2001) Effects of habitat fragmentation on choke disease (*Epichloe bromicola*) in the grass *Bromus erectus*. J. Ecol. 89:247–55

Holt, R. D. (1996) Food webs in space: an island biogeographic perspective. Pages 313 323 in G.A. Polis K.O. Winemiller, editors. Food webs: integration of patterns and dynamics. Chapman and Hall, London.

Kempson D., Lloyd M., Ghelardi R. (1963) A new extractor for woodland litter. Pedobiologia 3: 1-21.

Kruess, A., Tscharntke, T., (1994) Habitat fragmentation, species loss, and biological control. Science 264, 1581–1584.

Laurance, W.F. and Bierregaard, R.O. (1996) Fragmented Tropical Forests. Bulletin of the Ecological Society of America 77:34-36.

Lawrence K.L., Wise D.H. (2000) Spiders predation on forest-floor Collembola and evidence for indirect effects on decomposition. Pedobiologia 44:33-39

Lawrence K.L., Wise D.H. (2004) Unexpected indirect effect of spiders on the rate of litter disappearance in a deciduous forest. Pedobiologia 48:149-151

Nentwig W (1986) Non-webbuilding spiders: prey specialists or generalists. Oecologia 69:571-576

Nyffler M. (1999) Prey selection of spiders in the field. Journal of Arachnology 27:317-324.

Riechert, S.E. and Bishop, L. (1990) Prey control by an assemblage of generalist predators: Spiders in a garden test systems. Ecology, 71, 1441–1450.

Riechert, S.E. and Lawrence, K. (1997) Test for predation effects of single versus multiple species of generalist predators: spiders and their insect prey. Entomologia Experimentalis et Applicata, 84, 147–155.

Sanders, D. & Platner, C. (2007) Intraguild interactions between spiders and ants and top-down control in a dry grassland. Oecologia, 150, 611–624.

Saunders, D.A., Hobbs, R.J., Margules, C.R. (1991) Biological consequences of ecosystem fragmentation: a review. Conservation Biology 5, 18–32.

Schauermann, J. (1982) Verbesserte Extraktion der terrestrischen Bodenfauna im Vielfachgerät modifiziert nach Kempson und MacFadyen. Kurzmitteilungen aus dem SFB 135 (Ökosysteme auf Kalkgestein) 1:47-50.

Simberloff, D. (2000) What do we really know about habitat fragmentation. –Texas J. Sci. 52 (Suppl.): 5–22.

Snyder, W.E., Snyder, G.B., Finke, L.F. & Straub, C.S. (2006) Predator biodiversity strengthens herbivore suppression. *Ecology Letters*, **9**, 789–796.

Wise D.H. (2004) Wandering spiders limit densities of a major microbi-detritivore in the forest-floor food web. Pedobiologia 48:181-188.

Wise, D.H. (1993) Spiders in ecological webs. Cambridge University Press, Cambridge.

Zschokke, S., Dolt, C., Rusterholz, H.-P., Oggier, P., Braschler, B., Thommen, G.H., Lüdin, E., Erhardt, A., Baur, B. (2000) Short-term responses of plants and invertebrates to experimental small-scale grassland fragmentation. Oecologia, 125, 559–572.

Chapter 7

Chapter 7

Potential positive effect of the ant species *Lasius niger* on linyphiid spiders

Sebastian Schuch, Christian Platner and Dirk Sanders

Journal of Applied Entomology 2008,132, 375-381

Abstract

Ants are highly abundant generalist predators and important ecosystem engineers which can strongly affect the composition of animal comunities. We manipulated the density of the ant species *Lasius niger* with baits in a small scale field experiment to study the role of intraguild predation, top-down control and bottom-up effects of ants in a dry grassland surrounded by agriculture. Two different kinds of baits (honey and tuna) were exposed near six *L. niger* colonies and at a distance of two metres to the colonies in a dry grassland, where *L. niger* was a highly abundant, omnipresent species. The experiments were performed for one month in spring. Additionally, the natural abundance of *L. niger* varying with the distance to their nests was used to study the effects on spiders and potential prey groups.

The activity of *L. niger* was significantly higher at tuna baits compared to honey baits and empty control dishes. We found no effects of higher activity of *L. niger* on the arthropod community. However, there is evidence for a facilitation effect of ants on Collembola near to their colonies, probably due to habitat modification, which also influenced the density of Linyphiidae. Both groups had up to four times higher denisties next to *L. niger* colonies than at a distance of two meters. Furthermore the $\delta^{13}C$ values demonstrated that linyphiid spiders and *L. niger* predominantly feed on Collembola.

We conclude, that there is no evidence of top-down-effects of *L. niger* in a grassland in spring, however we found a facilitation of linyphiid spiders and their prey driven by the ants as ecosystem engineers.

Keywords: baits, Collembola, spring, small scale, facilitation, indirect effect, grassland

Introduction

Ants are one of the most influential groups in terrestrial ecosystems (Hölldobler & Wilson 1990) and are important as ecosystem engineers (Folgarait 1998). They influence the abiotic as well as the biotic environment. It is known that ants modify chemical and physical soil properties. For example, a higher total N-concentration occurs in eight-year-old mounds of prairie ants as compared to surrounding soil (Lane & BassiriRad 2005). Another study showed that the pH-value in *L. niger* colonies was decreased in nests built in acidic ground and increased in ant nests built in basic soil (Frouz et al. 2003). Furthermore, particle size and bulk density in colonies of *Lasius flavus* were lower in parts of the nest than in surrounding soil (Dostál et al. 2004). Nkem et al. (2000) demonstrated that soil modified by presence of ants had less clay components, a higher percentage of sand and silt, and a lower content of Ca, Mg, K and Na in comparison to the surrounding soil. Dauber and Wolters (1999) found higher microbial activity in mounds of *L. niger* compared to ground around the mounds.

Some studies of ants deal with top-down effects and intraguild predation (i.e. predation in the same guild). An ant removal experiment of Laakso and Setälä (1997) showed, that the absence of forest ants reduced the biomass of lumbricid earthworms. In contrast, the biomass of predatory arthropods caught in pitfall traps was increased. Kajak et al. (1972) detected a strong influence of *Myrmica* on herbivorous invertebrates in meadows. Moja-Laraño and Wise (2007) found that the reduction of wolf spider density by ants caused a higher density of Collembola (top-down effect). Also, Lonoir (2003) and Sanders and Platner (2007) found a negative effect of ants on web-building spiders.

In our study, we used two approaches: We manipulated the density of *L. niger* in a small scale field experiment at two distances from their colonies (experimental part) by exposing honey and tuna baits. And we measured the natural density of ants in the field (descriptive part) in order to detect the influence of ants on the arthropod community. We additionally used the analysis of stable isotopes to investigate trophic relationships between ants and their potential prey. We expected to find top-down effects or hints for intraguild predation.

Material and methods

Study site

The experiment was conducted in spring on a dry grassland near Göttingen (Lower Saxony, Germany) situated on a hill with an area of 50 x 140 metres surrounded by agricultural land. The research area was a Dauco-Picridetum hieracioidis within the Dauco-Melilotion association. Character species of this plant association is *Daucus carota* ssp. *carota* (wild carrot). *Picris hieracioides* ssp. *hieracioides* (hawkweed oxtongue), character species of the Dauco-Picridetum hieracioidis, was extremely dominant in the area. The average density of *L. niger* was 322 individuals/m² outside their colonies in early spring. In contrast, less than one individual/m² of

Formica was found. The epigeic Collembolans had an average density of 4500 individuals/m². The average density of web-building spiders (185 indiviuals/m²) and wandering spiders (143 individuals/m²) were quite similar.

Experiment and sampling

Six *L. niger* colonies were randomly selected from all nests in the dry grassland. To manipulate the density of *L. niger*, four different kinds of baits were used in the field: honey, tuna, honey/tuna and no food (O-samples). To protect the tuna and honey baits from vertebrates and weather influences, they were placed into petri dishes provided with eight entrance holes (8 mm Ø) on the border. The petri dishes were fixed to the ground by iron hooks. Each type of loaded petri dish was exposed directly next to a *L. niger* colony (0 m) and in a distance of two metres from the colony (2 m) resulting in 48 petri dishes for the six experimental units. The baits were controlled every second day for one month (22 visits) in spring 2006 (31st March 2006 to 4th April 2006). The baits were replaced every second visit. The activity of *L. niger* and other ants was recorded by counting the ants in the dish. After the observation period soil cores of 0.036 m² and 0.05 m depth were taken around each petri dish and fauna was extracted by heat (Kempson 1963; Schauermann 1982). Additionally, samples in 0 m and 2 m distance were taken from each *L. niger* colony at non-manipulated ground as controls (C). Samples were taken at the beginning and at the end of the study.

Furthermore, at the beginning of the study three additional colonies of *L. niger* were randomly selected. At these colonies samples were taken with a suction sampler (Stihl SH 85, Germany: 2x10s/sample using a 0.036 m² sampling cylinder) next to the colonies and at a distance of two metres. We used this additional sampling method to record the density of hypergeic arthropods. Most of the fauna was determined to family level. Collembolans were divided into endogeic and hypergeic individuals by size.

Stable isotopes

Ratios of ^{13}C and ^{15}N were estimated by a coupled system consisting of an elemental analyzer (Carlo Erba NA 2500) and a gas isotope mass spectrometer (Finnigan Delta^{plus}). The system is computer-controlled allowing measurement of ^{13}C and ^{15}N (Reineking et al. 1993). Isotopic contents were expressed in δ-units as the relative difference between sample and conventional standards with $δ^{15}$N or $δ^{13}$C [‰] = ($R_{Sample} - R_{Standard}$)/$R_{Standard}$ x 1000, where R is the ratio of ^{15}N/^{14}N or ^{13}C/^{12}C content, respectively. The conventional standard for ^{15}N is atmospheric nitrogen and for ^{13}C PD-belemnite (PDB) carbonate (Ponsard & Arditi 2000). Acetanilide (C_8H_9NO, Merck, Darmstadt) served for internal calibration with a mean standard deviation of samples <0.1‰. Samples were dried for 72 h (60°C) and weighed into tin capsules to contain 500-1800 µg of dry biomass. We analysed Linyphiidae, *L. niger* and *Formica rufibarbis* as predators and Collembola and Delphacidae (Auchenorrhyncha) as potential prey.

Data analysis

The effects of colony distance on ant activity and the density of ants, spiders and other arthropod groups were analysed by an analysis of variance (ANOVA). Effects of bait type on activity and density of these arthropod groups were analysed by a three-factor analysis of variance. All abundance data of the bait experiments were log-transformed to meet assumptions of normality and homogeneity of variance. Stable isotope data were analysed by performing a general linear model (GLM) due to the different sizes of the samples. All statistical analysis were performed with SAS (ver. 8: SAS, Cary, N.C.).

Results

Activity and density of *L. niger*

L. niger was up to two times more frequent at honey/tuna baits than at tuna baits whereas density was three times lower at sugar baits compared to honey/tuna baits. The empty control petri dishes were only rarely visited by ants (Table 1, Fig. 1). These effects were observed close to and 2 m away from *L. niger* colonies. The density of *L. niger*, as measured from soil samples, showed no significant pattern, except for a higher abundance at baits compared to the control (without any petri dishes; cf. Table 1) in two metres distance.

Tab. 1 Response of ant activity to treatment (22 visits within one month), abundance of *L. niger* (soil samples) and abundance of linyphiid spiders (soil samples), using a three-factor ANOVA. Differences between empty petri dishes and O-samples were tested using a two-factor ANOVA. All data were log transformed (log 10X + 1).

		Distance		Sugar		Thuna		Sugar*Thuna			Petri dish*O-sample	
	df	F	P	F	P	F	P	F	p	df	F	p
L. niger activity	1;35	0.09	0.761	159.01	<0.0001**	382.90	<0.0001**	1115	<0.0001**	-	-	-
Linyphiidae	1;35	16.08	0.0003**	0.16	0.692	0.01	0.926	0.01	0.915	1;15	6.98	0.019*
Lasius niger	1;35	1,55	0,222	0,30	0,590	2,31	0,138	0,67	0,418	1;15	0,62	0,751

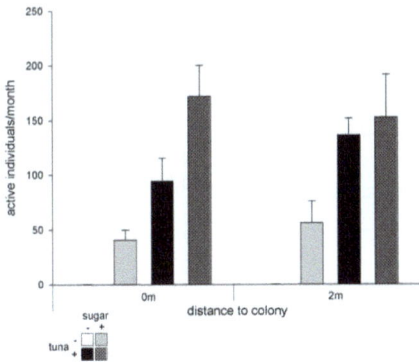

Fig. 1 Activity of *L. niger* (±SE) at baits in specially prepared petri dishes next to their colonies and in a distance of two metres. All activities of foraging ants at one type of bait are summed up (22 visits). Ant activity was measured from 29th March to 3rd May 2006 every second day. For statistical analysis see Table 1.

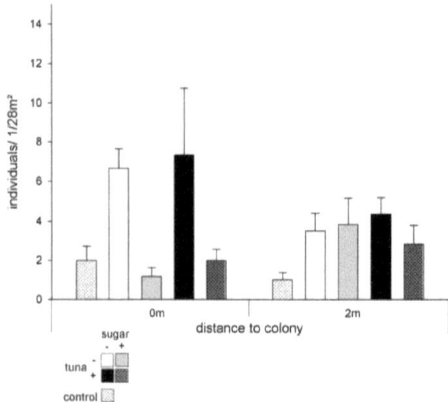

Fig. 2 Average density per 1/28 m² (±SE) of linyphiid spiders in different distances from colonies of L. niger (0 m and 2 m). The soil samples were taken around different kinds of bait in spring and treated by heat extraction. For statistical analysis see Table 1.

Abundance of other arthropod groups

Linyphiidae in bait plots were two times more abundant close to colonies of L. niger than in two metres distance (cf. Table 1, Fig. 2). Additionally, in samples from plots with empty petri dishes linyphiids were two to three times more abundant than in samples without dishes (cf. Table 1). There were no significant differences in the density of Linyphiidae within all kinds of baits. The results of the suction samples showed that Linyphiidae were up to eight times more abundant near colonies of L. niger than in the two-metre-distance (Table 2, Fig. 3B). Collembola densities also showed significant patterns in the suction samples in the same way as linyphiids (cf. Table 2, Fig. 3C) and a tendency to higher abundance in non-manipulated earth core samples taken in early spring (cf. Table 2, Fig. 3C). Both epigeic and hypergeic individuals of springtails were twice as abundant near the colonies as compared to two metres away from colonies of L. niger (cf. Table 2, Fig. 3C). The abundance of L. niger showed no significant pattern in suction samples (cf. Table 2, Fig. 3A).

The following groups did not respond to manipulation: Aphidinae, Lithobiidae, Geophilidae, Diplopoda, Nematocera, Brachycera, Symphyta, Apocrita, Isopoda, Carabidae, Thysanoptera, Araneidae, Hahniidae, Liocranidae, Lycosidae, Salticidae.

Table 2 Abundance data from suction and soil core samples of *Lasius niger*, linyphiid spiders and Collembolans in different distances from L. niger colonies (0 m and 2 m). Differences were tested using an ANOVA. $*p<0.05$, indicating statistical significance

Sampling method	Suction			Soil		
Dependent	Distance			Distance		
variable	FG	F	p	FG	F	p
Lasius niger	1;4	1.03	0.37	1;10	0.82	0.39
Linyphiidae	1;4	10.81	0.03*	1;10	2.50	0.15
Collembola total	1;4	12.46	0.02*	1;10	4.22	0.07
Collembola small	1;4	18.15	0.01*	1;10	4.07	0.07
Collembola large	1;4	5.28	0.08	1;10	2.22	0.17

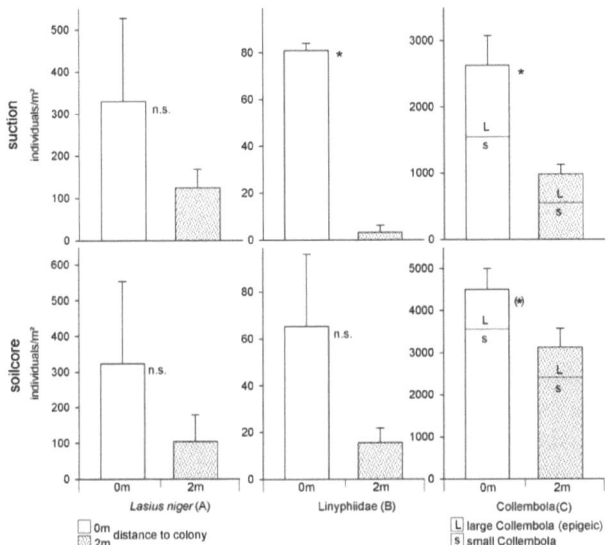

Fig. 3 Average density per m² (±SE) of *Lasius niger*, linyphiid spiders and Collembolans in different distances from *L. niger* colonies (0 m and 2 m). All suction and soil samples were taken in early spring (29.03.07). Significance is given between 0 m and 2 m values. For statistical data see Table 2.

Stable isotope measurement

The values of $\delta^{15}N$ and $\delta^{13}C$ for Delphacidae were 0.25‰ and -28.9‰, respectively (Tab. 3, Fig. 4). For *L. niger* and linyphiid spiders we found very similar $\delta^{15}N$ (3.5‰), ($F_{1;9}=1.16$, $p=0.31$, for GLM) and $\delta^{13}C$ (–27.5‰ and –27.7‰) values ($F_{1;9}=0.45$, $p=0.52$, for GLM). For Collembola $\delta^{13}C$ values (–27.8‰) are similar to the values of *L. niger* and Linyphiidae ($F_{2;13}=0.32$, $p=0.731$, for GLM), but the $\delta^{15}N$ values are –0.8‰ and differ significantly from those of the predators ($F_{2;13}=23.38$, $p<0.0001$, for GLM). Individuals of *F. rufibarbis* were more enriched in $\delta^{15}N$ and $\delta^{13}C$ values than other predators (4.7‰ ,$F_{1;22}=4.71$, $p=0.041$, for GLM; –25.6‰, $F_{1;22}=22.81$, $p<0.0001$, for GLM).

Discussion

The manipulation of *L. niger* activity was successful. The baits influenced the activity of *L. niger* in different ways: Tuna and honey/tuna baits were preferred most by *L. niger* workers; empty bait dishes were hardly visited. The colony requires more protein

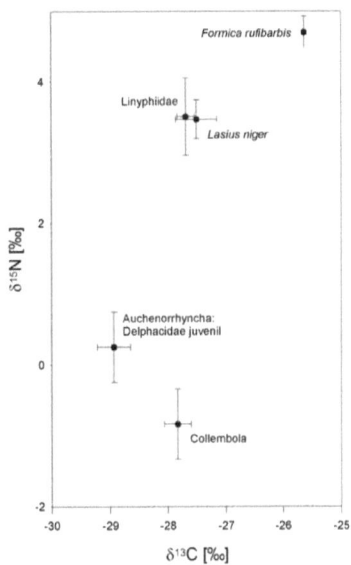

Fig. 4 $\delta^{15}N$ and $\delta^{13}C$ values (±SE) of Linyphiidae (6), Collembola (4), Delphacidae (6), *L. niger* (6) and *Formica rufibarbis* (3) sampled in early spring. Number of samples analyzed are given in parantheses.

during the beginning of the vegetation period, perhaps because of the so-called slow brood which outlasts the winter and is raised in spring (Hölldobler and Wilson 1990). Kajak et al. (1972) found a positive correlation between larvae produced by members of *Myrmica* and the amount of prey caught. Additionally, they found that total consumption by ants is generally highest in the first half of the growing season. However, our data from soil core samples taken around the baits showed no significant effects of a higher activity of *L. niger* on arthropod community in spring (data not presented).

Members of the family Linyphiidae were significantly more abundant close to colonies of *L. niger* than further away in both suction (Fig. 3B) and soil core samples taken around the baits (Fig. 2). There are three possible factors which could be responsible for higher densities of linyphiid spiders next to *L. niger* colonies: (1) more suitable microclimatic conditions, (2) a better habitat structure near mounds or (3) an increased food availability caused by a higher abundance of Collembola (Fig. 3C). Sparse vegetation on the mounds could lead to higher temperatures in spring due to more exposure to sunlight and a heating-up of the dark soil. Linyphiid spiders could profit from warmed air near the ground. The different habitat structure near ant colonies caused by boundary layer between herb and naked mound soil could also positively influence linyphiid spiders. Both factors may play a role in influencing the density of Linyphiid spiders, but such data are missing. It can be assumed that higher densities of Collembolans lead to higher densities of linyphiid spiders. Agustí et al. (2003) found that epigeic Collembola serve as an important food source of Linyphiidae in arable ecosystems. Furthermore, Alderweireldt (1994) found Collembola being 29.8% of total prey captured by linyphiid spiders in agricultural fields. Marcussen et al. (1999) emphasized that the importance of Collembolans for linyphiid spiders seems to be generally accepted. Wise et al. (1999) pointed out that theory predicts that spiders are preying on important detritivores and fungivores. Collembola as a fungivore group may find better conditions near *L. niger* colonies caused by the activity of the ant workers as ecosystem engineers. The $\delta^{15}N$ values of Collembolans are 3,5-4‰ lower than those of *L. niger* and members of Linyphiidae. That is an additional hint of Collembola being preyed upon both predator groups. Obviously, workers of *L. niger* do not feed regularly on Linyphiid spiders.

In conclusion, we found no evidence for top-down effects or intraguild predation of *L. niger*. However, we assume that the activity of *L. niger* in the soil caused higher Collembola densities next to colonies and affected through this bottom-up effect positively the density of linyphiid spiders, which profit from a higher prey density.

Acknowledgments

We thank Matthias Schaefer, Jordi Moya-Laraño and one anonymous referee for valuable discussion and comments on the manuscript. We are grateful to Sharon Cooper for linguistic corrections. Roswitha Schmickl provided important information on the plant association of the research area.

References

Alderweireldt M, 1994. Prey selection and prey capture strategies of linyphiid spiders in high-input agricultural fields. Bulletin of the British Arachnological Society 9, 300-308.

Agustí N, Shayler J, Harwood D, Vaughan P, Sunderland K, Symondson W, 2003. Collembola as an alternative prey sustaining spiders in arable ecosystems: prey detection within predators using melocular markers. Molecular Ecology 12, 3467-3475.

Dauber J, Wolters V, 1999. Microbial activity and functional diversity in the mounds of three different ant species. Soil Biology & Biochemistry 32, 93-99.

Dostál P, Březnová M, Kozlíčková V, Herbena T, Kovář P, 2005. Ant-induced soil modification and its effect on plant below-ground biomass. Pedobiologia 49, 127-137.

Folgarait PJ, 1998. Ant biodiversity and its relationship to ecosystem functioning: a review. Biodiversity and Conservation 7, 1221-1244.

Frouz J, Holec M., Kalčík J, 2003. The effect of *Lasius niger* (Hymenoptera, Formicidae) ant nest on selected soil chemical properties. Pedobiologia 47, 205-212.

Hölldobler B, Wilson EO, 1990. The Ants. 1st edition. Springer-Verlag, Berlin Heidelberg New York.

Kajak A, Breymeyer A, Petal J, Olechowicz E, 1972. The influence of ants on the meadow invertebrates. Ekologia Polska 17, pp. 163-170.

Kempson D, Lloyd M, Ghelardi R, 1963. A new extractor for woodland litter. Pedobiologia 3, 1-21.

Laakso J, Setälä H, 1997. Nest mounds of red wood ants (*Formica aquilonia*): hot spots for litter-dwelling earthworms. Oecologia 111, 565-569.

Lane DR, BassiriRad H, 2005. Diminishing effects of ant mounds on soil heterogeneity across a chronosequence of prairie restoration sites. Pedobiologia 49, 359-366.

Marcussen B, Axelsen J, Toft S, 1999. The value of two Collembola species as food for a linyphiid spider. Entomologia Experimentalis et Applicata 92, 29-36.

Nkem, J.N., Lobry de Bruyn, L.A., Grant, C.D. and N.R. Hulugale (2000) The impact of ant bioturbation and foraging activities on surrounding soil properties, Pedobiologia 44, 609–621.

Moya-Laraño J, Wise DH, (2007). Direct and indirect effects of ants on a forest-floor food web. Ecology, in press.

Reineking A, Langel R, Schikowski J, 1993. 15N, 13C-on-line measurements with an elemental analyzer (Carlo Erba, NA 1500), a modified trapping box and a gas isotope mass spectrometre (Finnigan, MAT 251). Isotopenpraxis 29:169-174.

Sanders D, Platner C, (2007). Intraguild interactions between spiders and ants and top-down control in a grassland food web. Oecologia 150:611-624.

Schauermann J, 1982. Verbesserte Extraktion der terrestrischen Bodenfauna im Vielfachgerät, modifiziert nach Kempson und MacFadyen. Kurzmitteilungen aus der SFB135: Ökosysteme auf Kalkstein 1, pp. 47-50.

Wise D, Snyder W, Tuntibunpakul P, Halaj J, 1999. Spiders in decomposition food websof agroecosystems: theory and evidence. The Journal of Arachnology 27, 363-370.

Chapter 8

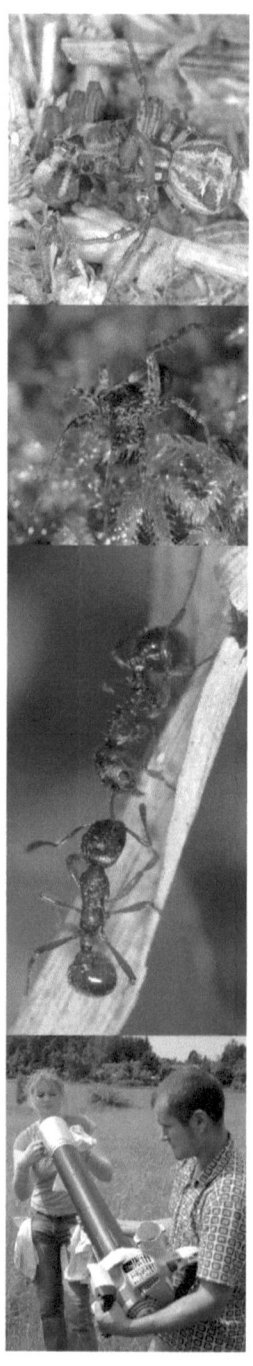

Chapter 8

Conclusion

The aim of this study with its several field experiments was to detect trophic interactions such as top-down and bottom-up effects and intraguild interference in grassland food webs. I focussed on spiders and ants as most abundant generalist predators in this grassland systems. The experiments were conducted in different habitats, ranging from dry grasslands, fallows, meadows to wet grasslands. On the one hand this diversity of habitat types makes comparisons difficult, however, on the other hand there is the advantage of identifying similarities and differences of food webs. The occurrence of top-down effects in terrestrial ecosystems is evident as demonstrated in some of my field experiments and many other studies. However, it is not sufficient to refer only to the publications in ecological journals, because many studies that failed to demonstrate strong top-down effects were probably not published. Another aim of the study was to reveal mechanisms that influence the strength of top-down forces. If these forces are often harder to detect than expected, what are the underlying mechanisms? This will be discussed by a comparison of field experiments in different habitats and the results from experiments with a manipulation of factors such as habitat structure and the diversity of the predator guild.

Food web interactions of generalist predators

In contrast to spiders as predators, most **ant** species are **omnivores**, being able to prey on a wide range of other invertebrates (Kajak *et al.* 1972), as well as to take up nutrients from plants indirectly by trophobiosis with phloem-feeding insects (Seifert 2007). Most ant species which nest in the ground throw up large mounds of soil and debris, thus changing soil properties and habitat heterogeneity by **bioturbation** (Dauber and Wolters 1999, Nkem et al. 2000, Frouz et al. 2003, Dostál et al. 2004). This may explain the occurrence of positive and negative effects of ants on prey populations and members of the same guild in some of the field experiments (Fig. 1).

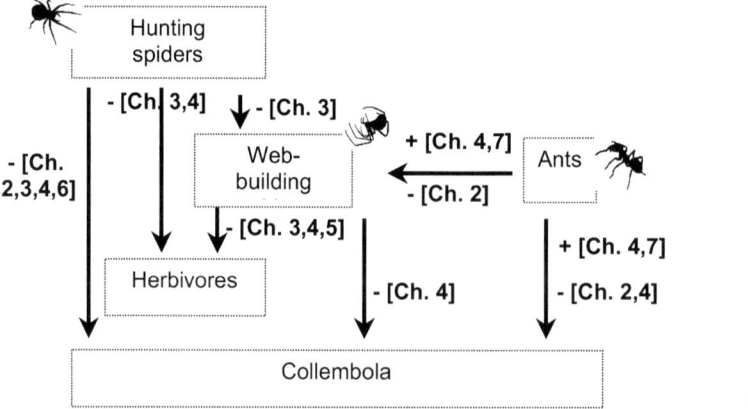

Fig. 1 Model of trophic interactions in the studied grassland food webs. Interactions can be positive or negative, which results in a change of the population size of the interacting group. The number of the chapters that demonstrated this effects is given in parentheses.

mbola may profit from the bioturbation by ants and positive effects of ants on web-building spiders are possibly based on this effect which indirectly affects the spider population positively by providing a more abundant prey group (Fig. 2).

The analysis of the stable isotope content of ^{14}N /^{15}N revealed that the **top predators** in the grasslands were **hunting spiders** such as *Pardosa* and *Pisaura*. Hunting spiders generally had higher $\delta^{15}N$ values than web-building spiders and ants indicating that intraguild predation is common among these spiders, which is also supported by their similar $\delta^{15}N$ values compared to the "spider eater" *Ero cambridgei* (Kulczynski) (see Chapter 4). Most of the field experiments demonstrated strong **top-down effects** of hunting spiders on Collembola, a trophic interaction, which was also found for web-building spiders and ants (cf. Fig. 1). Collembola are a high abundant arthropod group in the grassland ecosystems and seem to be the most important prey group for the three generalist predator groups. Especially juvenile spiders depend on Collembola as the main food resource, which was demonstrated by similar $\delta^{13}C$ values of Collembola and many ground living spider species and by $^{15}N/^{14}N$ ratios just one trophic level above those of Collembola.

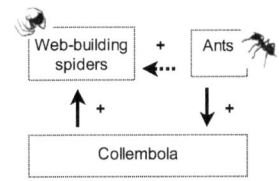

Fig. 2 Model of a possible indirect interaction between ants and web-building spiders, which is based on the density of Collembola.

Top-down effects on herbivores were strong in habitats which consist mainly of **one plant species**, i.e. the grass species *Agropyron repens* (L.), *Brachypodium pinnatum* (L.) and *Phalaris arundinacea* L., while for the dry grasslands with a **mix of plant species** no evidence for strong top-down effects on herbivores were found. Probably because of the diverse plant community the herbivore guild was also a diverse mix of different insect groups resulting in an alleviated top down control. The most abundant herbivores in the studied grasslands were planthoppers and leafhoppers (Auchenorrhyncha). In the wet grassland only one abundant Auchenorrhyncha species was affected while two other important herbivores showed no response to predator removal. These effects were observed for hunting spiders and web-builders in contrast to ants, which had no effect on herbivore populations at all. The predator assemblage containing the **wasp spider** *Argiope bruennichi* (Scopoli) strongly affected the most abundant planthoppers and leafhoppers and increased the diversity of this prey group. The occurrence of top down control in the presence of spiders and ants is not self-evident, because all prey groups are r-selected with a huge amount of offspring, and additionally abiotic factors can play an important role in dampening the top-down effects.

Intraguild interference on community level was found for ants and hunting spiders affecting the web-builders negatively, whereas no interactions occurred between ants and hunting spiders. The presence of hunting spiders reduced the strength of top-down effects of web-builders and ants caused a decrease in population size of web-building spiders. During my field work, I visually observed all kinds of intraguild predation between the three groups in the field with the exception of web-builders feeding on hunting spiders. Intraguild predation seem to be a common phenomenon in natural predator assemblages, but was detected on community level only in some cases (Chapter 2 and 3).

Top-down control of generalist predators on herbivores and detritivores in natural grassland systems

I studied the influence of **habitat structure** and **diversity of the predator guild** on the top-down control of generalist predators. Habitat structure and in particular vegetation height and architectural complexity strongly modified and indirectly affected the diversity of herbivorous arthropods (Fig. 3). I found that simply-structured habitats enhanced plant- and leafhopper suppression by spiders and ants. This effect was explained by a change in the composition of the predator assemblage with changing habitat structure and by a higher number of refuges for prey in more complex habitats. However, complex vegetation may also promote strong top-down effects by reducing antagonistic interactions among predators (Finke and Denno 2002; Corkum and Cronin 2004). The **fragmentation experiment** revealed a positive effect of smaller patch sizes on spider diversity and density of hunting spiders and a reduced density of Collembola in these smaller patches, which may be an indication of a stronger top-down control. **Intraguild interactions** are known to reduce the strength of top-down forces (Rosenheim et al. 1993, Lang 2003, Arim and Marquet 2004, Finke and Denno 2003, 2004, 2005), which was also demonstrated in a field experiment with web-builders and hunting spiders. By including intraguild predators such as hunting spiders into the predator guild, the *per capita* effects and top-down effects on Auchenorrhyncha population declined with increasing predator abundance and diversity. Similarly, predatory effects by hunting spiders and web-builders on a planthopper species and Collembola were reduced in the presence of ants, indicating intraguild interactions, that reduced the strength of top-down forces.

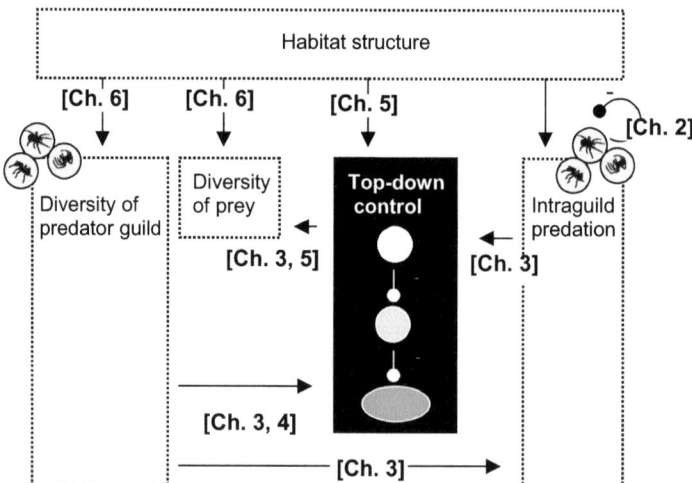

Fig. 3 Model of factors influencing the strength of top-down forces. The number of chapters that support this interaction is given in parentheses.

In a predator-removal experiment, the total number of Auchenorrhyncha species and the Shannon-Wiener index of the Auchenorrhyncha community were positively correlated with predator presence. One possible explanation for this effect in our experiment is release from competitive pressure from the two dominant leafhoppers species, which had lower densities in the presence of predators. In another experiment the species diversity and the density of Auchenorrhyncha declined in the presence of spiders, indicating that the less abundant herbivore guild contained also fewer species. To conclude, top-down effects were affected by habitat structure, predator identity and diversity and depend also on the community structure at lower trophic levels. If the communities contain many species belonging to different functional groups, top-down effects are harder to detect.

Short answers to the questions I addressed in the introduction

(1) Top-down control of generalist predator could be demonstrated in some of the field experiments and I found that the population density of Collembola was affected most frequently.

(2) Intraguild predation within a predator guild containing spiders and ants seemed to be common and important for ecosystem processes.

(3) A more diverse predator guild exerted a stronger top-down control on herbivores, however, if intraguild predators are included the strength of this interaction decreased.

(4) Top down effects attenuated with increasing structural complexity of the habitat by providing refuges for herbivores and by changing the composition of the predator assemblage.

(5) Small scale fragmentation affected species diversity and abundance of generalist predators positively.

Field experiments having a high external validity (Naem 2001) are an excellent method for studying trophic interactions and predatory effects under natural conditions. However, manipulation in the field is difficult and a combination with mesocosm experiments under more controlled conditions as done by Finke and Denno (2005) can provide important information concerning the mechanisms of effects. In addition to the field experiments I used stable isotopes to get information of the trophic structure of the food webs. All these approaches are a very fruitful combination of methods for understanding trophic linkages in natural food webs.

For future studies, it would be interesting to focus on the importance of abundant and of rare species for ecosystem processes, because rare species are most likely to get extinct and their function for ecosystem processes is mostly unclear.

References

Arim, M. & Marquet, P.A. (2004) Intraguild predation: a widespread interaction related to species biology. Ecology Letters, 7, 557–564.

Corkum, L..D., & Cronin, D.J. (2004) Habitat complexity reduces aggression and enhances consumption in crayfish. Journal of Ethology, 22, 23-27.

Dauber, J., Wolters, V. (1999) Microbial activity and functional diversity in the mounds of three different ant species. Soil Biology & Biochemistry 32, 93-99.

Dostál, P., Březnová, M., Kozlíčková, V., Herbena, T., Kovář ,P. (2005) Ant-induced soil modification and its effect on plant below-ground biomass. Pedobiologia 49, 127-137.

Finke, D.L., & Denno, R.F. (2002) Intraguild predation diminished in complex-structured vegetation: Implications for prey suppression. Ecology, 83, 643-652.

Finke, D.L. & Denno, R.F. (2003) Intra-guild predation relaxes natural enemy impacts on herbivore populations. Ecological Entomology, 28, 67–73.

Finke, D.L. & Denno, R.F. (2004) Predator diversity dampens trophic cascades. Nature, 429, 407–410.

Finke, D.L. & Denno, R.F. (2005) Predator diversity and the functioning of ecosystems: the role of intraguild predation in dampening trophic cascades. Ecology Letters, 8, 1299–1306.

Frouz, J., Holec, M., Kalčík, J. (2003) The effect of *Lasius niger* (Hymenoptera, Formicidae) ant nest on selected soil chemical properties. Pedobiologia 47, 205-212.

Kajak, A., Breymeyer, A., Pêtal, J., & Olechowicz, E. (1972) The influence of ants on the meadow invertebrates. Ekologia Polska, 20 (17), 163-171.

Lang, A. (2003) Intraguild interference and biocontrol effects of generalist predators in a winter wheat field. Oecologia, 134,144–153.

Naem, S. (2001) in Gardner, R. H., Kemp, W.M., Kennedy, V.S., and Petersen, J. E., editors. (2001) Scaling relations in experimental ecology. Columbia University Press, New York, New York, USA.

Nkem, J.N., Lobry de Bruyn, L.A., Grant, C.D. and Hulugale, N.R. (2000) The impact of ant bioturbation and foraging activities on surrounding soil properties, Pedobiologia 44, 609–621.

Rosenheim, J.A., Wilhoit, L.R. & Armet, C.A. (1993) Influence of intraguild predation among generalist insect predators on the suppression of herbivore population. Oecologia, 96, 439–449.

Seifert, B. (2007) Die Ameisen Mittel- und Nordeuropas. Lutra Verlag, Trauer.

Danksagung

Mein herzlicher Dank gilt **Matthias Schaefer** für die Ausbildung in Ökologie und die Betreuung der Arbeit. Besonders die breite Formenkenntnis, die auf den Exkursionen in Deutschland und am Gardasee vermittelt wurde, ist neben einem fundierten Theoriegebäude mit vielen Fallbeispielen eine wichtige Grundlage ökologischen Verständnisses.

Auch **Christian Platner** gilt mein Dank hinsichtlich der Betreuung der Arbeit, besonders in der Anfangsphase. Er war immer offen für Projektideen und Diskussionen über Freilandexperimente und Statistik und hat sich auch vor der schweißtreibenden Arbeit im Freiland keineswegs gescheut.

Klaus Hövemeyer danke ich für die erhellenden Gespräche über Diversitätsindizes.

Werner und **Waltraut Sanders** möchte ich für das Fundament an ökologischem Wissen und die fortwährende Unterstützung danken. Verbundenheit mit der Natur sind die Grundlage für Interesse und Motivation welche wiederum die Schlüsselfaktoren zum Gelingen einer Arbeit sind.

Herbert Nickel war außerordentlich hilfreich beim Bestimmen von Zikaden und beim Beantworten von Fragen bezüglich ihrer Biologie.

Ohne Hilfe von **Heiko Bischoff, Simone König, Sebastian Schuch** und **Christina Taraschewski**, beim Sortieren der Proben und zusammen mit **Thomas Grützner, Karl Winkelgrund** und **Dieter Nünchert** bei der Freilandarbeit mit der zum Teil doch recht anstrengenden Probenahmen, wären viele Experimente nicht möglich gewesen. Nicht zu vergessen sind die vielen vergnüglichen Stunden, da Freilandarbeit im Team einfach Spaß macht!

Ebenfalls beim Bearbeiten der Proben halfen **Christel Fischer, Ingrid Kleinhans, Renate Grüneberg** und **Susanne Böning-Klein**.

Ganz besonderes danke ich **Sharon Cooper,** die an vielen Bereichen der Arbeit beteiligt war: Freiland, Probenahme und immer wieder mit dem Lesen von Manuskripten beschäftigt

Curriculum vitae

Appendix

Appendix 1 Phytosociological record of the five different blocks in July 2002. The five blocks were located along the hillside with each in different vegetation, % for cover of single species.

Block (each consists of 4 plots)	1	2	3	4	5
	Dry grassland			Meadow	
Sanguisorba minor Scop.	2				
Corylus avellana L.	5				
Agropyron repens L.	2				
Fragaria vesca L.	2	2			
Leucanthemum vulgare (Lam.)	2	5	20		
Lotus corniculatus L.	6	1	5		
Silene vulgaris Garcke		2			
Medicago lupulina L.	10	4	10	20	10
Clinopodium vulgare L.	17	50	40	15	15
Euphorbia cyparissias L.	2	10			5
Viola hirta L.	4	10		5	10
Hypericum perforatum L.	2	1		4	4
Poa pratensis L.	7	2	2		4
Bromus erectus Huds.	2			5	
Agrimonia eupatoria L.	15	5	5	4	20
Centaurea jacea L.	1	4		2	
Knautia arvensis (L.) Coult	2	2	4		2
Trisetum flavescens (L.) P.Beauv.	1	3	4	2	2
Cirsium arvense Scop.	2				4
Arrhenatherum elatius (L.) P.Beauv.		2	5	5	2
Brachypodium pinnatum (L.) P.Beauv.		1	5	30	
Anthriscus sylvestris Hoffm.		4	5	2	5
Tanacetum vulgare L.			2		
Astragalus spec.			5	20	
Galium album Mill.				5	2
Height of herb layer (cm)	70	100	120	120	120
Height of shrub layer (cm)	50				110
Cover of herbs (%)	60	85	85	95	85
Cover of mosses (%)	30	60	70	40	60
Total vegetation cover (%)	80	100	95	100	100
Cover of litter (%)		10	50	60	60

Appendix 2 Arthropod taxa referred to in Figure 4, with values and standard deviations of δ ^{13}C and δ ^{15}N. * = juvenile web-building spiders (Linyphiidae, Theridiidae, Tetragnathidae, Araneidae), n = number of samples for analysis

	δ ^{13}C	δ ^{15}N	SD δ^{13}C	SD δ^{15}N	n
Wandering spiders					
Alopecosa trabalis (Clerck)	-27,1	2,5	0,5	0,6	6
Arctosa lutetiana (Simon)	-25,7	4,4	0,2	0,1	3
Aulonia albimana (Walckenaer)	-28,3	0,9	0,5	0,8	30
Clubiona juv	-27,4	-0,1	0,4	0,2	5
Lycosidae	-27,0	3,1	0,8	1,0	12
Pardosa juv	-27,3	1,9	0,8	0,7	5
Pardosa lugubris (Walckenaer)	-27,7	4,1			1
Pisaura juv	-26,9	-0,5			1
Pisaura mirabilis (Clerck)	-26,3	2,8	0,2	0,1	2
Tibellus juv	-27,0	-0,2			1
Tibellus oblongus (Walckenaer)	-28,2	3,3			1
Trochosa terricola Thorell	-27,9	2,8	0,1	0,4	1
Zora sylvestris Kulczynski	-27,0	0,9	0,5	0,2	1
Web-building spiders					
Argiope bruennichi (Scopoli)	-26,1	4,4			1
Atypus piceus (Sulzer)	-25,1	3,4	0,5	0,1	3
Mangora acalypha (Walckenaer)	-26,6	4,8			1
Meioneta rurestris (C.L.Koch)	-28,0	-0,4			1
Tenuiphantes tenuis (Blackwall)	-28,4	1,1	0,7	1,2	4
Walckenaeria acuminata Blackwall	-27,7	3,4			1
Web-builders	-27,9	1,4	0,3	1,5	9
Web-builders juv*	-27,7	1,5	0,4	0,3	3
Ants					
Formica cunicularia Latreille	-26,9	2,2	0,2	0,2	2
Lasius alienus Förster	-28,0	0,7	1,3	0,6	6
Lasius flavus (Fabricius)	-27,5	0,4	1,0	0,3	5
Myrmica sabuleti Meinert	-26,8	1,3	0,1	0,5	5
Ponera coarctata Latreille	-28,2	2,7			1
Herbivores and detritivores					
Aphidinae a	-29,0	-5,4			1
Aphidinae b	-26,5	-0,3			1
Auchenorrhyncha group a	-29,1	-3,6	0,7	0,8	4
Arboridia parvula (Boheman)	-29,4	-2,5			1
Mocydiopsis attenuata (Germar)	-28,4	-4,1			1
Auchenorrhyncha group b	-25,5	-3,3	0,3	0,9	7
Adarrus multinotatus (Boheman)	-25,4	-4,0			1
Anoscopus albifrons (Linnaeus)	-25,4	-3,6			1
Aphrophora alni (Fallén)	-25,8	-2,3	0,1	0,1	3
Chrysomelidae (Alticinae)	-29,5	-2,2	0,5	0,2	3
Collembola	-27,8	-1,8	0,2	1,4	8
Diptera (Sphaeroceridae)	-28,0	5,7	0,2	0,1	3
Isopoda	-27,3	-1,1			1
Julidae	-25,9	-3,0	1,4	0,7	3
Ortheziidae	-29,9	-4,0	0,1	0,2	2
Plants					
Grass	-28,6	-4,2	0,1	0,8	6
Herb	-29,2	-3,5	0,21	0,4	6
Moss	-29,8	-4,8		0,4	60,5

Appendix 3 Dominance of leafhoppers in response to predator removal and vegetation height (uncut = 20 cm and cut = 10 cm).

Species	Uncut Vegetation		Cut		Predator effect ANOVA
	Removal	Predator	Removal	Predator	
Ribautodelphax pungens (Rib.)	46.1	24.0	35.4	28.7	*
Recilia coronifer (Marsh.)	26.0	11.0	40.4	30.0	*
Mocydia crocea (H.-S.)	9.7	13.2	14.8	13.8	ns
Stenocranus minutus (F.)	1.4	18.9		0.3	ns
Delphacodes venosus (Germ.)	1.4	11.9	1.4	0.9	ns
Delphacidae indet.	1.6	5.2		9.8	ns
Anoscopus flavostriatus (Don.)	3.7	3.4	1.5	4.9	ns
Hyledelphax elegantula (Boh.)	1.9	3.9	1.0	0.3	ns
Adarrus multinotatus (Boh.)	1.0	1.0	1.7	6.4	ns
Acanthodelphax spinosa (Fieb.)	3.9	0.3		0.3	ns
Anoscopus albifrons (L.)	1.9		1.5	1.5	ns
Errastunus ocellaris (Fall.)		2.9	0.7		ns
Arocephalus longiceps (Kbm.)			1.0	1.2	ns
Megophthalmus scanicus (Fall.)		1.1		0.3	ns
Deltocephalinae indet.	0.5	0.6			ns
Aphrodes cf. *bicincta* (Schrk.)	0.1	0.4		0.3	ns
Elymana sulphurella (Zett.)		0.4	0.2	0.3	(*)
Rhopalopyx preyssleri (H.-S.)	0.2	0.1	0.2	0.3	ns
Allygidius commutatus (Fieb.)	0.1	0.1		0.3	ns
Cicadula persimilis (Edw.)	0.4				-
Agallia spec.		0.3			-
Athysanus quadrum Boh.		0.3			-
Megamelus notula (Germ.)		0.3			-
Neophilaenus albipennis (F.)		0.1		0.3	-
Arthaldeus pascuellus (Fall.)		0.1			-
Athysanus argentarius Metc.		0.1			-
Euscelis incisus (Kbm.)	0.1				-
Javesella pellucida (F.)		0.1			-
Megadelphax sordidula (Stål)		0.1			-
Verdanus abdominalis (F.)			0.2		-
n total	**801**	**699**	**582**	**327**	

Wissenschaftlicher Buchverlag bietet
kostenfreie
Publikation
von
Dissertationen und Habilitationen

Sie verfügen über eine wissenschaftliche Abschlußarbeit zu aktuellen oder zeitlosen Fragestellungen, die hohen inhaltlichen und formalen Anspruchen genügt, und haben **Interesse an einer honorarvergüteten Publikation**?

Dann senden Sie bitte erste Informationen über Ihre Arbeit per Email an: info@svh-verlag.de.

Unser Außenlektorat meldet sich umgehend bei Ihnen.

Südwestdeutscher Verlag für Hochschulschriften
Aktiengesellschaft & Co. KG

Dudweiler Landstr. 99
D – 66123 Saarbrücken

www.svh-verlag.de

Printed by Books on Demand GmbH, Norderstedt / Germany